Springer Theses

Recognizing Outstanding Ph.D. Research

Springer Theses – the "best of the best"

Internationally top-ranked research institutes select their best thesis annually for publication in this series. Nominated and endorsed by two recognized specialists, each thesis is chosen for its scientific excellence and impact on research. For greater accessibility to non-specialists, the published versions include an extended introduction, as well as a foreword by the student's supervisor explaining the special relevance of the work for the field. As a whole, the series provides a valuable resource both for newcomers to the relevant field, and for other scientists seeking detailed background information on special questions. Finally, it provides an accredited documentation of the valuable contributions made by today's younger generation of scientists.

The content of the series is available to millions of readers worldwide and, in addition to profiting from this broad dissemination, the author of each thesis is rewarded with a cash prize equivalent to € 500.

More information about this series at http://www.springer.com/series/8790

Baran Sarac

Microstructure-Property Optimization in Metallic Glasses

Springer

Baran Sarac
Institute für Complex Materials
Leibniz Institute IFW Dresden
Dresden
Germany

ISSN 2190-5053 ISSN 2190-5061 (electronic)
Springer Theses
ISBN 978-3-319-36538-1 ISBN 978-3-319-13033-0 (eBook)
DOI 10.1007/978-3-319-13033-0

Springer Cham Heidelberg New York Dordrecht London

Softcover reprint of the hardcover 1st edition 2015

Printed on acid-free paper

Springer is part of Springer Science+Business Media (www.springer.com)

Foreword

This thesis consists of an in-depth study of investigating microstructure–property relationships in bulk metallic glasses using a novel quantitative approach by which influence of the second phase features on mechanical properties can be independently and systematically analyzed. We adopted this strategy to evaluate and optimize the elastic and plastic deformation, as well as the overall toughness of cellular honeycombs under in-plane compression and porous heterostructures under uniaxial tension. The first study revealed three major deformation zones in cellular metallic glass structures, where deformation changes from collective buckling showing nonlinear elasticity to localized failure exhibiting a brittle-like deformation, and finally to global sudden failure with negligible plasticity as the length to thickness ratio of the ligaments increases. For the second case, it has been found that spacing and size of the pores, the pore configuration within the matrix, and the overall width of the sample determines the extent of deformation, where the optimized values are attained for pore diameter to spacing ratio of one with AB-type pore stacking.

In general, this versatile concept can also provide an insight to a wide range of problems including very complicated microstructural architectures, stochastic foam designs found in nature, as well as flaw tolerance and sensitivity studies of different classes of materials.

Yale University, Department of
Mechanical Engineering and Materials Science

Prof. Jan Schroers
(Supervisor of Dr. Baran Sarac)

Acknowledgements

I take this opportunity to extend my gratitude to the people who have been instrumental in the successful completion of this thesis. I would like to express my deepest appreciation to my advisor, Prof. Jan Schroers, who has been abundantly helpful and has offered invaluable assistance and support. He is not only a successful scientist, but also a role model for the young scientists of our generation with his energy and dedication to his work. I will forever value our scientific discussions, and will always be inspired by his vision and perception of life.

I take immense pleasure in thanking Prof. Corey S. O'Hern and Prof. Aaron M. Dollar for serving on my advisory committee and providing me guidance and feedback on my thesis research. I would like to show my greatest appreciation to my external committee member, Prof. Robert D. Conner. I am also grateful to my former committee member, Prof. Ainissa G. Ramirez for her contributions and support in my special investigations and my area exam. I would also like to express my gratitude to Prof. John B. Morrell who acted as a committee member in my dissertation progress presentations.

Many thanks to our former postdoc, Dr. Golden Kumar, who has recently become a professor at Texas Technical University, for his vast contributions and his insightful comments about my research throughout my PhD.

I would like to thank Prof. Jamie Guest at Johns Hopkins University, who utilized computational algorithm to structurally optimize cellular structures. Special thanks to Dr. Amish Desai for providing me silicon templates and contributing to the revision of my paper about blow-molding of metallic glasses. I also thank A.J. Barnes, who sent me superplastically formable aluminum alloy pieces to generate heterostructures.

I would like to thank all of the undergraduate researchers who have worked/been working in my research projects and providing me experimental data. I owe thanks to my current and previous colleagues, friends, as well as many other people in our department and school, for their valuable ideas, support and assistance throughout my PhD study.

Finally, yet importantly, I would like to express my heartfelt thanks to my beloved parents, Nes'e Sarac and Prof. A. Sezai Sarac, for their blessings and guidance.

Article Note (Parts of this Thesis)

Parts of this thesis have been published in the following journal articles:

- **Sarac, B.**, Schroers, J., "Designing Tensile Ductility in Metallic Glasses", Nature ***Communications,*** 06/2013, vol. 4, pp. 1–7, DOI: 10.1038/ncomms3158.
- **Sarac, B.**, Schroers, J., "From brittle to ductile: Density optimization of Zr-based bulk metallic glass cellular structures", ***Scripta Materialia***, 06/2013, vol. 68, pp. 921–924, DOI: 10.1016/j.scriptamat.2013.02.030.
- **Sarac, B.**, Ketkaew, J., Popnoe, D. O. and Schroers, J., "Honeycomb Structures of Bulk Metallic Glasses", ***Advanced Functional Materials***, 04/2012, vol. 22, pp. 3161–3169, DOI: 10.1002/ adfm.201200539.
- **Sarac, B.**, Kumar, G., Hodges, T., Ding, S., Desai, A., and Schroers, J., "Three-Dimensional Shell Fabrication Using Blow Molding of Bulk Metallic Glass", ***Journal of Microelectromechanical Systems***, 02/2011, vol. 20, pp. 28–36, DOI: 10.1109/JMEMS.2010.2090495.

Contents

About the Author

Dr. Baran Sarac received his B.S. degree in metallurgical and materials engineering and mechanical engineering from Middle East Technical University, Ankara, Turkey. He has completed his masters and doctorate degree in the Department of Mechanical Engineering and Materials Science at Yale University, New Haven, CT, under the mentorship of Prof. Jan Schroers. He worked successively as a postdoctorate researcher in Helmholtz Zentrum Geesthacht for one year, and has recently embarked on his new position at Leibniz Institute, IFW Dresden with the same title on mechanical and functional characterization of intelligent alloy systems. His other research interests include structural design, thermoplastic forming, in-situ testing and morphological characterization of advanced cellular structures, as well as numerical simulations of superplastic materials via finite element analysis.

Through his studies at Yale University, Dr. Sarac has been entitled to several esteemed awards, including 2013 Yale University Harding Bliss Prize owing to his contributions to further the intellectual life of the Yale School of Engineering & Applied Science, Pierre W. Hoge fellowship (between 2008-2009), and 2012 Materials Research Society Fall Best Poster Award. His publications have appeared in peer reviewed international journals such as Nature Communications, Advanced Functional Materials, Acta Materialia, Materials Letters, Scripta Materialia, and Journal of Microelectromechanical systems (IEEE), where he was concomitantly involved in federal research projects of DARPA and US Department of Energy.

Chapter 1
General Introduction

1.1 Motivation and Scope of Complex Materials

The demands on materials increase rapidly. To meet these demands, more and more complex microstructures and microstructural architectures are used. However, currently used strategies to develop increasingly complex structures are unsuited to create tomorrow's materials. For example, the main challenge in determining microstructure–property relationships is that any kind of individual variation in feature properties inevitably changes other properties. This limitation is due to the fabrication methods, which do not permit completely independently vary just one microstructure feature. As an example, if one attempts to alter, for instance, the spacing of a phase in a microstructure, at the same time, length, volume, composition, dispersity, and density of this phase will also change because all the properties are interconnected [1, 2].

To address this challenge, computational methods have been exploited. Even though virtual experiments have been powerful in explaining specific aspects of microstructure–property relationships [3–7], they have limitations when considering "real microstructures." This is due to multiscale nature of complex microstructures, and often times, constitutive equations are unknown particularly for their plastic behavior. One example is molecular dynamics simulations which have been widely used to investigate structure–property relationships. Today's available computing power; however, limits the system size and simulation time, where the limited system size for molecular simulations cannot capture the multiple length scale microstructures. This limitation is even more obvious in ab-initio calculations, which typically permits calculations with up to ~100 atoms [8]. For example, if a representative microstructure is at least 10 μm^3, the analysis of this microstructure requires at least 10^{12} atoms, which is well beyond today's ab-initio capacities. Alternatively, model systems have also been employed to investigate and visualize microstructure–property relationships [9, 10]. Based on Bernal's hard sphere approach [11–13], colloidal systems were used [14, 15] to understand the interparticle interactions' role on the kinetics of crystallization and glass formation [16–18]. Even though colloidal approach is powerful in some aspects, the limitations of these

B. Sarac, *Microstructure-Property Optimization in Metallic Glasses*, Springer Theses,
DOI 10.1007/978-3-319-13033-0_1

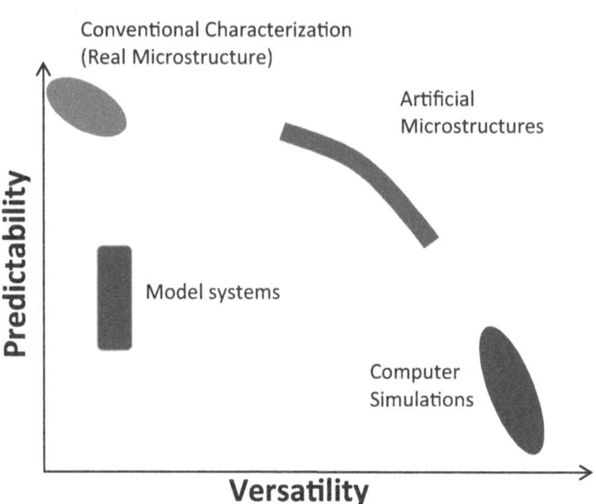

Fig. 1.1 Predictability versus versatility. Comparison of different approaches to determine microstructure–property relationship: Conventional characterization methods, computer simulations, model systems, and artificial microstructures. Within these approaches, artificial microstructures combine high predictability with high versatility

methods are only a small number of properties that can be varied and inaccurate representation of the forces.

Current fabrication methods limit the experimental characterization methods, resulting in low-structural versatility. For instance, commercial methods used for micropatterning cannot be easily applied to cellular or high-porosity materials due to low-lateral resolution in microlevel precision, as well as the reactivity of the master alloy with the cutting tools. In comparison, the feature variations can be done very easily with the simulation methods, but the results neither reflect nor predict the "real" microstructures with high accuracy. Especially for the molecular dynamic simulations which utilizes atomic interactions and accurately predicts the material behavior at nanoscale, the fundamental drawback arises as the entire sample size become orders of magnitude bigger than the characteristic feature size, causing inaccurate representation of atomic potentials and limiting the applicability of this technique on wider length scales. Artificial microstructure approach combines the favorable properties of both predictability and versatility via its unique fabrication strategy and quantification of results, making them excellent candidates to analyze complex microstructures (Fig. 1.1). The findings of the proposed work will thus be essential for future design of novel complex microstructures with optimized mechanical performance.

The focus of this thesis is to decode complex microstructures and establish microstructure–property relationships using our novel artificial microstructure approach. Two examples will be considered in detail: behavior of periodic cellular structures under in-plane compression and designing heterostructures for tensile ductility. Metallic glasses (MGs) were chosen due to their interesting and impressive suite of mechanical and physical properties. To synthesize artificial microstructures, desired patterns were designed and fabricated into silicon, and the patterns were subsequently transferred into the MG. Using this approach, the effect of individual microstructural feature (independent of other variables) on mechanical response and

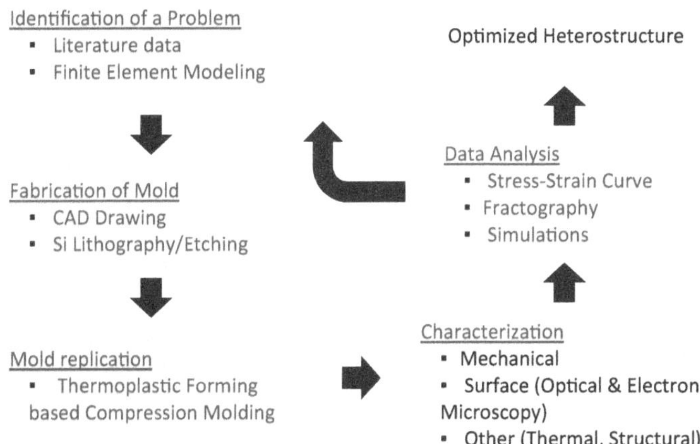

Fig. 1.2 Sketch of our proposed strategy to design, fabricate, and characterize heterostructures with individual control over microstructural architectures

toughening mechanism has been systematically and quantitatively analyzed under uniaxial tension, in-plane compression and bending at ambient temperatures and quasistatic loading conditions. Post-characterization and data analysis provides information for further design improvements. Figure 1.2 shows our strategy to design and characterize optimized microstructural architectures.

1.2 An Overview of Metallic Glasses (MGs)

MGs will be the focus material of this thesis because of their technological relevance [19–23], their thermoplastic forming ability (TPF) [24–29], and size effects on their mechanical properties [26, 30–36]. MGs are a new class of engineering materials which are multicomponent alloys that vitrify with remarkable ease during solidification [24]. They offer unique combinations of strength, ductility, toughness, and processability due to the absence of grain boundaries and dislocations in the glassy state. Particularly for structural applications, they possess a wide array of attractive properties (Fig. 1.3) including high (specific) strength and elasticity combined with good corrosion and wear resistance [37]. In addition, some MGs show comparable fracture toughness with steels (e.g., Fe, Ti, and Zr-based MGs; Fig. 1.4), which is a measure of mechanical viability [38].

In addition, recent advancements in MG processing allow net-shaping of MGs into geometries that were previously unachievable with any other metal forming process [25, 29, 39–41]. This combination of properties and processability promises a range of potential applications for MGs. In particular during thermoplastic forming, MGs can be considered high-strength structural metals that can be processed like plastics [42]. The homogeneous and isotropic nature of their amorphous

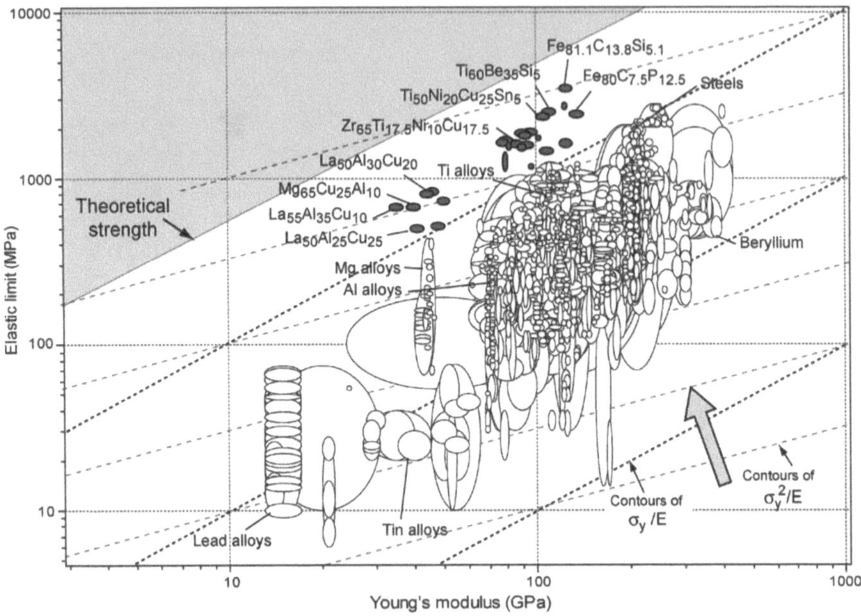

Fig. 1.3 Ashby map showing elastic limit σ_y plotted against modulus E for various metals, alloys, metal matrix composites, and metallic glasses (MGs). Yield strain σ_y/E and resilience σ^2_y/E are defined by the contour lines. MGs exhibit very high strength, much closer to the theoretical limit than their crystalline counterparts, and elasticity often in combination with high corrosion and wear resistance. (Reprinted from reference [50] with a permission from Elsevier)

structure opens new possible application fields in micro- and nanoscales [25, 26, 29, 43–49]. The properties of MGs are summarized in Table 1.1, where they are categorized by pros and cons.

A major drawback of MGs is the lack of tensile ductility. This is due to an absence of a strain-hardening mechanism, which is the origin of tensile ductility in crystalline metals. Instead, a strain-softening mechanism results in instability by localizing shear into narrow bands. Such shear bands result in catastrophic failure often along a single shear band.

Although it has been shown that even though typical MGs lack any tensile ductility in bulk form, in geometries where one dimension approaches 1 mm, significant bending ductility is observed [32]. In this thesis, we attempt to exploit the aforementioned attractive properties of MGs while improving weaknesses such as toughness and ductility using artificial microstructures. Required for this strategy is an understanding of the underlying mechanistic origins of MG heterostructures.

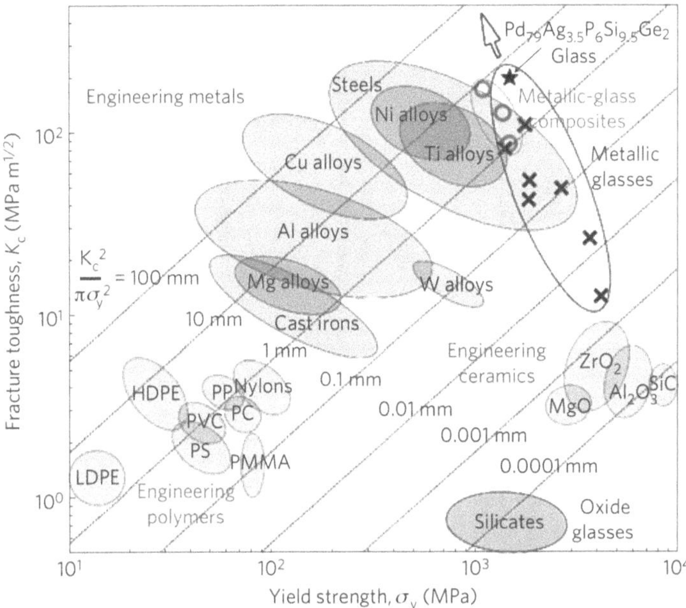

Fig. 1.4 Metallic glasses (MGs) are known to be among the strongest engineering materials together with a large variation in fracture toughness, which is the material's ability to resist against cracks. (Reprinted from reference [38] with a permission from Nature Publishing group)

1.3 Processing of MGs

MGs can be thermoplastically formed, which is unique among metals [39]. In the processing of MGs, the metastable nature of their amorphous state imposes the main challenge. Any fabrication process must avoid crystallization during processing and solidification since the desired MG properties are lost after crystallization, and the crystallized MG former can no longer be thermoplastically formed [24]. Crystallization occurs when the cooling or heating path intersects with the crystallization curve of the time–temperature-transformation (TTT) diagram (Fig. 1.5; [24]). This intersection can be prevented by two principally different processing paths. Path 1 indicates the critical cooling rate for glass formation, which is the slowest cooling rate that avoids crystallization. Thus, cooling rates faster than path 1 must be achieved during casting to avoid the "crystallization nose." In direct casting (path 1), the MG former must be cooled faster than the critical cooling rate to avoid crystallization and simultaneously fill the entire mold cavity during solidification. These contradicting requirements make the fabrication of thin sections with a high aspect ratio particularly challenging. Furthermore, the high temperatures involved in direct casting make this process incompatible with the current fabrication techniques [42].

Table 1.1 Properties of metallic glasses (MGs) in different application fields. The absence of grain boundaries and dislocations in glassy alloys contributes to the exceptional combination of mechanical, magnetic, chemical, and tribological properties (Reprinted from reference [50] with a permission from Elsevier)

Attributes	Attractive attributes	Unattractive
General	Absence of microstructural features such as grain and phase boundaries and of related composition variations (e.g., segregation). This allows components with features of near-atomic scale	Present cost of components and processing Optimization of composition for glass-forming ability prevents easy optimization for other properties, including low density
Mechanical	High hardness, H, giving good wear and abrasion resistance High yield strength, cr_yFracture toughness K_c and toughness G_c can be very high High specific strength[a], σ_y / ρ, $\sigma_y^{2/3} / \rho$, and $\sigma_y^{1/2} / \rho$ High resilience per unit volume and mass[a], σ_y^2 / E and $\sigma_y^2 / E\rho$ Low-mechanical damping	Severe localization of plastic flow (shear-banding), giving zero ductility in tension Fracture toughness K_c and toughness G_c can be very low Can be embrittled by annealing Small process-zone size ($d < 1$ mm) means that larger components may fail in a brittle manner
Thermal	$T_g < T_c$ for some MGs, allowing processing as a supercooled liquid (T_g—glass-transition temp. T_c—temp of crystallization onset)	Instability above T_c limits high temperature use
Electrical and magnetic	High-magnetic permeability Resistivity is nearly independent of temperature	Relatively high magneto-striction gives energy loss in oscillating field
Chemical	Lack of grain structure and associated microstructural features (e.g., solute segregation) gives corrosion resistance	
Environmental	Some compositions biocompatible	Not easily recycled once in a product (nonconventional compositions)
Processing	Low-solidification shrinkage and lack of grain structure give high precision and finish in castings The high viscosity and low-strain-rate sensitivity of the supercooled liquid permit thermoplastic forming	Current need for vacuum die-casting gives relatively slow production rate
Aesthetic	Lack of grain structure allows a very high polish High hardness and corrosion resistance gives durability	
Potential markets	Aesthetics, present novelty and rarity make MGs attractive for high-end "life-style" products Properties and processing favour μm-to-mm scale structures	Current high cost of material and processing limits applications to those with high value-added

[a] Material indices of this kind are discussed in Ref. [14]

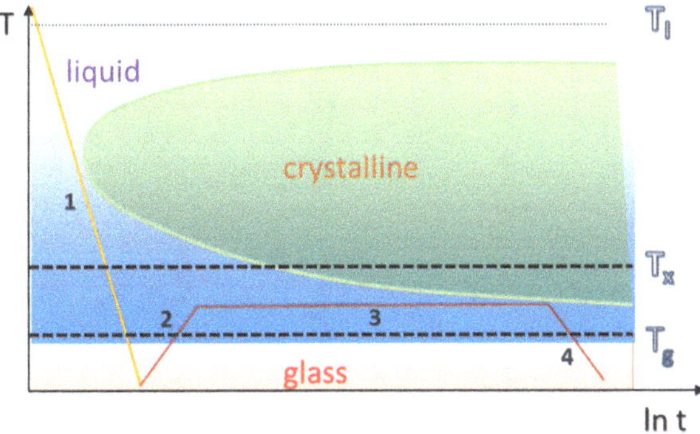

Fig. 1.5 Schematic TTT diagram illustrating the processing methods of MG formers, where *T* is temperature, *t* is time, T_l is liquid temperature, and T_g is glass transition temperature. During direct casting (path 1), the liquid is filled into mold and simultaneously cooled fast to avoid crystallization. During TPF (path 2–4), the required fast cooling and shaping are decoupled. In order to prepare the feedstock material for TPF, the amorphous MG is cast into simple shapes such as rods, bars, discs, or plates. Subsequently, this feedstock material is reheated into the supercooled liquid region where the processing window widens and the formation of complex geometries can be better controlled. After the TPF process, no fast cooling is required to avoid crystallization during cooling

Path 2–4 represents the processing paths for TPF [24]. In this process, the preparation of an amorphous structure and the forming are decoupled. The forming takes place when the MG is reheated into the supercooled liquid region (path 2), where the amorphous phase relaxes into a viscous liquid [51] and flows under low applied pressure before it eventually crystallizes. As the processing time window available in path 3 is much longer than during direct casting and the processing temperature is significantly lower, a better control over the process and higher dimensional accuracy can be achieved. No fast cooling is required to avoid crystallization after the forming step during TPF because of sluggish crystallization kinetics (path 4). We have recently shown that, as long as crystallization is avoided during TPF-based processing, and the critical fictive temperature of the MG is lower than its glass transition [52], the mechanical properties of the MG are unaffected [53]. The accessible processing viscosities of 10^6 Pa·s and temperatures for some MGs, ~160 °C for $Au_{49}Ag_{5.5}Pd_{2.3}Cu_{26.9}Si_{16.3}$ and ~270 °C for $Pt_{57.5}Cu_{14.7}Ni_{5.3}P_{22.5}$, are comparable to plastics [54–57].

Figure 1.6 compares the flow stress curves of different materials with respect to temperature change. The ideal processing region is defined by the lowest processing pressure where effects such as turbulent flow and gravitational influences can be neglected on the time scale of the experiment. This region can be reached by plastics, which has contributed to the wide proliferation of plastics in industrial societies over the past 60 years. However, no conventional metals or even superplastically formable (SPF) alloys exhibit a comparable processing behavior. The main drawback of plastics in structural applications is that they are weak. Metals, on the contrary, exhibit high-strength values, but are difficult to process because of the

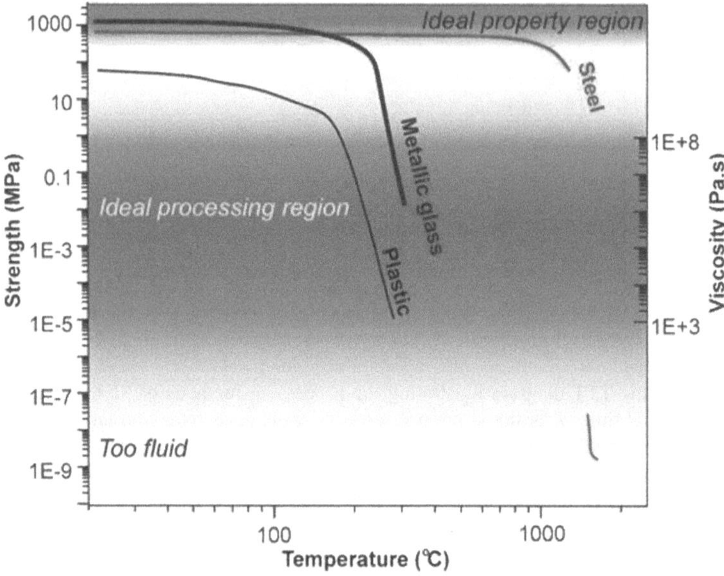

Fig. 1.6 MGs combine two mutually exclusive attributes when they are thermoplastically formed: the strength of high-strength metals, and the processability of plastics. (Reprinted from reference [59] with a permission from Elsevier)

abrupt change in viscosity at their melting temperatures. Compared to plastics and high-strength steels, highly processable MGs [54, 55, 57, 58] exhibit room temperature strength, that is, two orders of magnitude higher than plastics and 2–3 times higher than conventional steels. Thus, MGs combine the strength of metals with the processability of plastics [59]. Figure 1.7 shows the wide range of length scales and geometries that can be fabricated with TPF-based processes [19, 30, 60]. Covering nine orders of magnitude in length scale (from ångström (Å) to meter (m)), MGs can be utilized for versatility of shapes, i.e., from atomically smooth surface roughness up to meter long strings [26].

MG properties have been found to be depending on size (Fig. 1.8). Samples with an effective thickness (smallest characteristic dimension) exceeding 1–5 mm are often brittle in bending and tension but show some compressive plasticity [61–64]. Multiple shear bands are activated during compression of a monolithic MG, particularly in alloys with a low ratio of shear modulus to bulk modulus, which dramatically increases the plasticity of the sample [62, 65]. When MGs are used in geometries where one dimension is below about ten times its critical crack length (~ 1 mm for a medium range Zr-based MG like Vit1), they exhibit significant bending plasticity [32, 66]. It has been observed [67, 68] and explained experimentally that shear band confinement changes the deformation mode [35] as long as the spacing of the second phases matches the plastic zone size of the related MG matrix. A transition from shear localization (inhomogeneous) to homogeneous deformation was observed (in a temperature and strain rate region where deformation is typically localized

Fig. 1.7 Processing flexibility of MGs allow us to fabricate shapes at different length scales with high precision and roughness unachievable with other conventional alloys

into shear bands; [69]) when the size of the sample decreases below approximately 100 nm [26, 70]. Significant plasticity has also been observed in nanoscale MG heterostructures [63, 71]. However, spacing of the second phase is significantly (up to a factor of 50) smaller than the critical confinement length scale of ~ 100 nm. For these nanoscale heterostructures, the control over individual microstructure features is even more limited [35]. When the size of the sample matches the size of cluster of atoms (below ~ 10 nm), inelastic shear distortion can be expected, creating shear transformation zones (STZs). At this length scale, variation of length to thickness

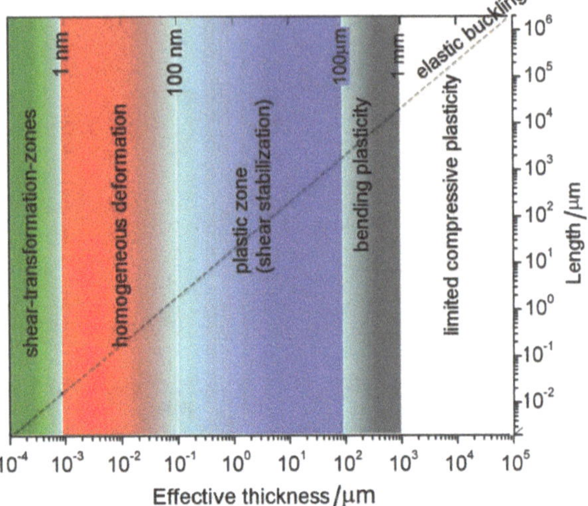

Fig. 1.8 Size effects in metallic glasses (MGs) estimated for $Zr_{41\cdot2}Ti_{13.8}Cu_{12.5}Ni_{10}Be_{22.5}$ (Vit1). (Reprinted from reference [26] with a permission from Elsevier)

of the sample sets another length scale, where the deformation mode can be altered from brittle to ductile, which results in elastic buckling prior to reaching σ_y [26].

1.4 Mechanical Property Enhancement in MG Composites

It has been earlier reported that the transformation of a shear band into a crack is hampered by MGs with a high Poisson's ratio (or *B*/*G*, the ratio between the bulk modulus, *B*, and the shear modulus, *G*; [62, 64, 65]). However, this trend has only been observed when samples were characterized in compression or bending geometry but not under tensile loading.

Other strategies have been developed to encumber the immediate transformation of a shear band into a crack. Most effective has been the introduction of the second phase microstructures into the MG. Such MG heterostructures can be realized through partial crystallization [34, 72–75], chemical decomposition [76], forming a composite by mixing with a second phase [77–79], and by incorporation of a gas phase to fabricate cellular materials and foams [80–84].

Plasticity of MGs can be improved by any means of increasing shear band density in the MG resulting in a better distribution of the deformation energy within the sample [85]. It has been proposed that the interaction between shear bands and second phases plays a critical role in stabilizing shear bands and preventing crack formation [15–20]. For example, it was suggested that shear bands could be stabilized as long as the spacing of the second phases is comparable with the plastic zone size (R_p) of the sample [15]. This includes multiplication, branching, and restriction of the shear bands, thus controlling the instabilities otherwise responsible for early failure. Therefore, R_p was suggested to be a crucial length scale influencing shear band interaction with the second phase. More specifically, when the spacing of the second phases is lower than the plastic zone size, shear bands will not immediately develop into cracks, but instead, the formation of multiple shear bands is triggered [36].

The shear band density can be increased by hindering the propagation of shear bands, and at the same time, aiding the nucleation of new shear bands. This can be achieved by any means of reducing the stress at the tip of the shear band, for instance by placing obstacles in their way. Shear bands can also be initiated intentionally by introducing sites of local stress concentration in the material. Effective initiation sites should be densely spread throughout the material and initiate shear bands at similar stress levels. Possible initiation sites are pores or second phase particles with a Young's modulus and ductility significantly different of the matrix material, leading to interfacial stresses. These effective obstacles should arrest shear band propagation or at least split up propagating shear bands [85]. Independently, the length of the processing zone can also be varied by considering a range of MGs with varying intrinsic toughness [62, 86] and by inducing structural relaxation [87].

The most common fabrication method of MG heterostructures is either through a melt infiltration process [77, 88–91] or via extrusion using TPF process [92–94]. Some of these composites fabricated showed a minor increase in plasticity under compressive loading conditions (see, e.g., [78]). However, the limited enhancement of plasticity and particularly, the inability to stabilize shear band propagation under tensile loading was attributed to a chemical reaction between the MG matrix and the (second phase) second phases. This reaction typically results in formation of a brittle phase at the interface (see, e.g., [95]). The drawback of a weak interface can be eliminated by partially crystallizing the MG former. These so-called in situ composites or precipitate reinforced MGs (ex situ) typically consist of a ductile second phase which crystallizes upon cooling while the remaining melt freezes into a glass [34, 72–74, 96]. The properties of the second phases are essential for global ductility since they have to be able to absorb the strain of the penetrating shear band. It has been found that when a high-volume fraction of soft and ductile second phases is introduced into the MG through partial crystallization, significant ductility and enhancement in toughness can be achieved, and values of up to 10% for ductility and $K_{1c} = 170$ MPa m$^{-1/2}$ have been reported [34, 72, 97].

Different microstructural features contribute to the toughening of MG heterostructures. Recent studies focused on creating an inhomogeneous microstructure with isolated dendrites in a MG matrix stabilizes the glass against the catastrophic failure associated with unlimited extension of a shear band, which results in enhanced global plasticity and delay in fracture [34, 72, 98, 99]. Figure 1.9 depicts micrographs of some examples of dendritic MG composites, which are formed by β-phase stabilization (Fig. 1.9a, c, d) and semi-solid processing (Fig. 1.9b). These two-phase heterostructures show variation in spacing, size, and shape of the second phase, as well as the chemical composition of the second phase can also vary throughout the structure.

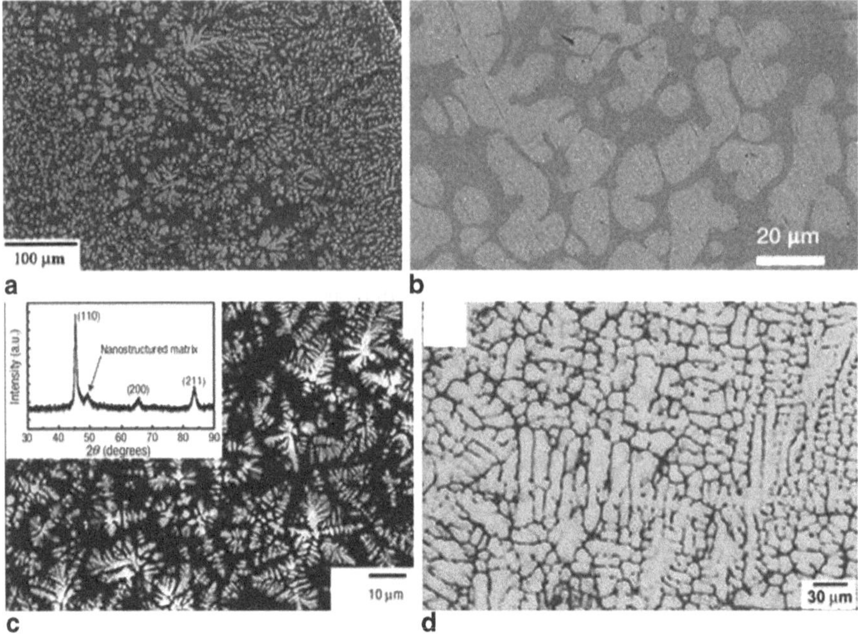

Fig. 1.9 Microstructures of MG heterostructures that exhibit tensile ductility. Variation in spacing, size, and shape of the second phase dendrites is present in each sample. **a** $(Zr_{75}Ti_{18.34}Nb_{6.66})_{75}X_{25}$, **b** $Zr_{39.6}Ti_{33.9}Nb_{7.6}Cu_{6.4}Be_{12.5}$, **c** $Ti_{60}Cu_{14}Ni_{12}Sn_{4}Ta_{10}$, **d** $Ti_{66.1}Cu_{8}Ni_{4.8}Sn_{7.2}Nb_{13.9}$ in situ composites

As it can be seen from this figure, MG composites have very complex microstructures, where broad ranges of contributions have been suggested for the toughening mechanism. To name the most prominent, this includes the softness (shear modulus) of the second phase, the strength and toughness of the interface, the morphology and volume fraction of the second phase. It should be mentioned that previous studies do "suggest" these contributions, but do not provide quantitative experimental evidence. Furthermore, with current MG heterostructure synthesis techniques, variation of microstructural features tends to be limited, and changes are often interrelated. The artificial microstructure approach introduced in this thesis targets to overcome these shortcomings by creating a systematic and quantitative analysis through independent variation of variables.

References

1. Schwarz M, Karma A, Eckler K, Herlach DM. Physical-Mechanism of Grain-Refinement in Solidification of Undercooled Melts. Phys Rev Lett. 1994;73:2940.
2. Schroers J, HollandMoritz D, Herlach DM, Grushko B, Urban K. Undercooling and solidification behaviour of a metastable decagonal quasicrystalline phase and crystalline phases in Al-Co. Mat Sci Eng a-Struct. 1997;226:990.
3. Falk ML, Langer JS, Pechenik L. Thermal effects in the shear-transformation-zone theory of amorphous plasticity: comparisons to metallic glass data. Phys Rev E. 2004;70:011507-1.

4. Yamakov V, Wolf D, Salazar M, Phillpot SR, Gleiter H. Length-scale effects in the nucleation of extended dislocations in nanocrystalline Al by molecular-dynamics simulation. Acta Mater. 2001;49:2713.
5. Yamakov V, Wolf D, Phillpot SR, Gleiter H. Grain-boundary diffusion creep in nanocrystalline palladium by molecular-dynamics simulation. Acta Mater. 2002;50:61.
6. Lund AC, Schuh CA. Strength asymmetry in nanocrystalline metals under multiaxial loading. Acta Mater. 2005;53:3193.
7. Li QK, Li M. Free volume evolution in metallic glasses subjected to mechanical deformation. Mater Trans. 2007;48:1816.
8. Hood RQ, Galli G. Insulator to metal transition in fluid deuterium. J Chem Phys. 2004;120:5691.
9. Torquato S, Stillinger FH. Jammed hard-particle packings: from Kepler to Bernal and beyond. Rev Mod Phys. 2010;82:2633.
10. Song C, Wang P, Makse HA. A phase diagram for jammed matter. Nature. 2008;453:629.
11. Bernal JD, Finney JL. Random close-packed hard-sphere model. 2. Geometry of random packing of hard spheres. Discuss Faraday Soc. 1967;43:62.
12. Bernal JD. Geometry of the structure of monatomic liquids. Nature. 1960;185:68.
13. Bernal JD, Mason J. Co-ordination of randomly packed spheres. Nature. 1960;188:910.
14. Pusey PN, Vanmegen W. Phase-behavior of concentrated suspensions of nearly hard colloidal spheres. Nature. 1986;320:340.
15. Vanblaaderen A, Wiltzius P. Real-space structure of colloidal hard-sphere glasses. Science. 1995;270:1177.
16. Royall CP, Poon WCK, Weeks ER. In search of colloidal hard spheres. Soft Matter. 2013;9:17.
17. Pusey PN, Zaccarelli E, Valeriani C, Sanz E, Poon WCK, Cates ME. Hard spheres: crystallization and glass formation. Philos T R Soc A. 2009;367:4993.
18. Schall P, Weitz DA, Spaepen F. Structural rearrangements that govern flow in colloidal glasses. Science. 2007;318:1895.
19. Schroers J. Bulk metallic glasses. Phys Today. 2013;66:32.
20. Chen MW. A brief overview of bulk metallic glasses. Npg Asia Mater. 2011;3:82.
21. Wang WH, Dong C, Shek CH. Bulk metallic glasses. Mat Sci Eng R. 2004;44:45.
22. Telford M. The case for bulk metallic glass. Mater. Today March. 2004;7:36.
23. Greer AL. Metallic glasses ... on the threshold. Mater Today. 2009;12:14.
24. Schroers J. The superplastic forming of bulk metallic glasses. Jom-Us. 2005;57:35.
25. Kumar G, Tang HX, Schroers J. Nanomoulding with amorphous metals. Nature. 2009;457:868.
26. Kumar G, Desai A, Schroers J. Bulk metallic glass: the smaller the better. Adv Mater. 2011;23:461.
27. Schroers J. On the formability of bulk metallic glass in its supercooled liquid state. Acta Mater. 2008;56:471.
28. Sarac B, Kumar G, Hodges T, Ding SY, Desai A, Schroers J. Three-dimensional shell fabrication using blow molding of bulk metallic glass. J Microelectromech S. 2011;20:28.
29. Schroers J, Pham Q, Desai A. Thermoplastic forming of bulk metallic glass—a technology for MEMS and microstructure fabrication. J Microelectromech S. 2007;16:240.
30. Sarac B, Ketkaew J, Popnoe DO, Schroers J. Honeycomb structures of bulk metallic glasses. Adv Funct Mater. 2012;22:3161.
31. Sarac B, Schroers J. From brittle to ductile: density optimization for Zr-BMG cellular structures. Scripta Mater. 2013;68:921.
32. Conner RD, Johnson WL, Paton NE, Nix WD. Shear bands and cracking of metallic glass plates in bending. J Appl Phys. 2003;94:904.
33. Wada T, Inoue A, Greer AL. Enhancement of room-temperature plasticity in a bulk metallic glass by finely dispersed porosity. Appl Phys Lett. 2005;86:251907-1.
34. Hofmann DC, Suh JY, Wiest A, Duan G, Lind ML, Demetriou MD, Johnson WL. Designing metallic glass matrix composites with high toughness and tensile ductility. Nature. 2008;451:1085.
35. Volkert CA, Donohue A, Spaepen F. Effect of sample size on deformation in amorphous metals. J Appl Phys. 2008;103:083539-1.

36. Sarac B, Schroers J. Designing tensile ductility in metallic glasses. Designing tensile ductility in metallic glasses. Nat Commun. 2013;4:2158-1.
37. Johnson WL. Bulk glass-forming metallic alloys: science and technology. Mrs Bull. 1999;24:42.
38. Demetriou MD, Launey ME, Garrett G, Schramm JP, Hofmann DC, Johnson WL, Ritchie RO. A damage-tolerant glass. Nat Mater. 2011;10:123.
39. Schroers J. Processing of bulk metallic glass. Adv Mater. 2010;22:1566.
40. Schroers J, Pham Q, Peker A, Paton N, Curtis RV. Blow molding of bulk metallic glass. Scripta Mater. 2007;57:341.
41. Martinez R, Kumar G, Schroers J. Hot rolling of bulk metallic glass in its supercooled liquid region. Scripta Mater. 2008;59:187.
42. Schroers J, Paton N. Amorphous metal alloys form like plastics. Adv Mater Process. 2006;164:61.
43. Saotome Y, Itoh K, Zhang T, Inoue A. Superplastic nanoforming of Pd-based amorphous alloy. Scripta Mater. 2001;44:1541.
44. Sharma P, Kaushik N, Kimura H, Saotome Y, Inoue A. Nano-fabrication with metallic glass—an exotic material for nano-electromechanical systems. Nanotechnology. 2007;18:035302-1.
45. Chu JP, Wijaya H, Wu CW, Tsai TR, Wei CS, Nieh TG, Wadsworth J. Nanoimprint of gratings on a bulk metallic glass. Appl Phys Lett. 2007;90:034101-1.
46. Kumar G, Schroers J. Write and erase mechanisms for bulk metallic glass. Appl Phys Lett. 2008;92:031901-1.
47. Pan CT, Wu TT, Chang YC, Huang JC. Experiment and simulation of hot embossing of a bulk metallic glass with low pressure and temperature. J Micromech Microeng. 2008;18:025010-1.
48. Inoue A, Nishiyama N. New bulk metallic glasses for applications as magnetic-sensing, chemical, and structural materials. Mrs Bull. 2007;32:651.
49. Henann DL, Srivastava V, Taylor HK, Hale MR, Hardt DE, Anand L. Metallic glasses: viable tool materials for the production of surface microstructures in amorphous polymers by micro-hot-embossing. J Micromech Microeng. 2009;19:115030-1.
50. Ashby MF, Greer AL. Metallic glasses as structural materials. Scripta Mater. 2006;54:321.
51. Busch R, Schroers J, Wang WH. Thermodynamics and kinetics of bulk metallic glass. Mrs Bull. 2007;32:620.
52. Kumar G, Neibecker P, Liu YH, Schroers J. Critical fictive temperature for plasticity in metallic glasses. Nat Commun. 2013;4:1536-1.
53. Kumar G, Rector D, Conner RD, Schroers J. Embrittlement of Zr-based bulk metallic glasses. Acta Mater. 2009;57:3572.
54. Schroers J, Lohwongwatana B, Johnson WL, Peker A. Gold based bulk metallic glass. Appl Phys Lett 2005;87:061912-1.
55. Zhang B, Zhao DQ, Pan MX, Wang WH, Greer AL. Amorphous metallic plastic. Phys Rev Lett. 2005;94:205502-1.
56. Legg BA, Schroers J, Busch R. Thermodynamics, kinetics, and crystallization of Pt57.3Cu14.6Ni5.3P22.8 bulk metallic glass. Acta Mater. 2007;55:1109.
57. Duan G, Wiest A, Lind ML, Li J, Rhim WK, Johnson WL. Bulk metallic glass with benchmark thermoplastic processability. Adv Mater. 2007;19:4272.
58. Schroers J, Johnson WL. Highly processable bulk metallic glass-forming alloys in the Pt-Co-Ni-Cu-P system. Appl Phys Lett. 2004;84:3666.
59. Schroers J, Hodges TM, Kumar G, Raman H, Barnes AJ, Quoc P, Waniuk TA. Thermoplastic blow molding of metals. Mater Today. 2011;14:14.
60. Carmo M, Sekol RC, Ding SY, Kumar G, Schroers J, Taylor AD. Bulk metallic glass nanowire architecture for electrochemical applications. Acs Nano. 2011;5:2979.
61. Xing LQ, Li Y, Ramesh KT, Li J, Hufnagel TC. Enhanced plastic strain in Zr-based bulk amorphous alloys. Phys Rev B. 2001;64:180201-1.
62. Schroers J, Johnson WL. Ductile bulk metallic glass. Phys Rev Lett. 2004;93:255506-1.
63. Das J, Tang MB, Kim KB, Theissmann R, Baier F, Wang WH, Eckert J. "Work-hardenable" ductile bulk metallic glass. Phys Rev Lett. 2005;94:205501-1.

64. Liu YH, Wang G, Wang RJ, Zhao DQ, Pan MX, Wang WH. Super plastic bulk metallic glasses at room temperature. Science. 2007;315:1385.
65. Lewandowski JJ, Wang WH, Greer AL. Intrinsic plasticity or brittleness of metallic glasses. Phil Mag Lett. 2005;85:77.
66. Schuh CA, Hufnagel TC, Ramamurty U. Overview No.144—mechanical behavior of amorphous alloys. Acta Mater. 2007;55:4067.
67. Schuh CA, Lund AC, Nieh TG. New regime of homogeneous flow in the deformation map of metallic glasses: elevated temperature nanoindentation experiments and mechanistic modeling. experiments Acta Mater. 2004;52:5879.
68. Guo H, Yan PF, Wang YB, Tan J, Zhang ZF, Sui ML, Ma E. Tensile ductility and necking of metallic glass. Nat Mater. 2007;6:735.
69. Spaepen F. Microscopic mechanism for steady-state inhomogeneous flow in metallic glasses. Acta Metallurgica. 1977;25:407.
70. Bharathula A, Lee SW, Wright WJ, Flores KM. Compression testing of metallic glass at small length scales: effects on deformation mode and stability. Acta Mater. 2010;58:5789.
71. Inoue A, Zhang W, Tsurui T, Yavari AR, Greer AL. Unusual room-temperature compressive plasticity in nanocrystal-toughened bulk copper-zirconium glass. Phil Mag Lett. 2005;85:221.
72. Hays CC, Kim CP, Johnson WL. Microstructure controlled shear band pattern formation and enhanced plasticity of bulk metallic glasses containing in situ formed ductile phase dendrite dispersions. Phys Rev Lett. 2000;84:2901.
73. Fan C, Ott RT, Hufnagel TC. Metallic glass matrix composite with precipitated ductile reinforcement. Appl Phys Lett. 2002;81:1020.
74. Kuhn U, Eckert J, Mattern N, Schultz L. ZrNbCuNiAl bulk metallic glass matrix composites containing dendritic bcc phase precipitates. Appl Phys Lett. 2002;80:2478.
75. Louzguine DV, Kato H, Inoue A. High-strength Cu-based crystal-glassy composite with enhanced ductility. Appl Phys Lett. 2004;84:1088.
76. Kundig AA, Ohnuma M, Ping DH, Ohkubo T, Hono K. In situ formed two-phase metallic glass with surface fractal microstructure. Acta Mater. 2004;52:2441.
77. Choi-Yim H, Johnson WL. Bulk metallic glass matrix composites. Appl Phys Lett. 1997;71:3808.
78. Szuecs F, Kim CP, Johnson WL. Title: Mechanical properties of Zr56.2Ti13.8Nb5.0Cu6.9Ni 5.6Be12.5 ductile phase reinforced bulk metallic glass composite. Acta Mater. 2001;49:1507.
79. Siegrist ME, Loffler JF. Bulk metallic glass-graphite composites. Scripta Mater. 2007;56:1079.
80. Schroers J, Veazey C, Johnson WL. Amorphous metallic foam. Appl Phys Lett. 2003;82:370.
81. Wada T, Inoue A. Formation of porous Pd-based bulk glassy alloys by a high hydrogen pressure melting-water quenching method and their mechanical properties. Mater Trans. 2004;45:2761.
82. Schroers J, Veazey C, Demetriou MD, Johnson WL. Synthesis method for amorphous metallic foam. J Appl Phys. 2004;96:7723.
83. Brothers AH, Dunand DC. Porous and foamed amorphous metals. Mrs Bull. 2007;32:639.
84. Brothers AH, Dunand DC. Ductile bulk metallic glass foams. Adv Mater. 2005;17:484.
85. Siegrist ME. Bulk metallic glass composites. Doctoral Thesis. 2007.
86. Hess PA, Poon SJ, Shiflet GJ, Dauskardt RH. Indentation fracture toughness of amorphous steel. J Mater Res. 2005;20:783.
87. Gilbert CJ, Ritchie RO, Johnson WL. Fracture toughness and fatigue-crack propagation in a Zr-Ti-Ni-Cu-Be bulk metallic glass. Appl Phys Lett. 1997;71:476.
88. Xing LQ, Herlach DM, Cornet M, Bertrand C, Dallas JP, Trichet MF, Chevalier JP. Mechanical properties of Zr57Ti5Al10Cu20Ni8 amorphous and partially nanocrystallized alloys. Mat Sci Eng a-Struct. 1997;226:874.
89. Xing LQ, Eckert J, Schultz L. Deformation mechanism of amorphous and partially crystallized alloys. Nanostruct Mater. 1999;12:503.
90. Conner RD, Dandliker RB, Scruggs V, Johnson WL. Dynamic deformation behavior of tungsten-fiber/metallic-glass matrix composites. Int J Impact Eng. 2000;24:435.
91. Fan C, Inoue A. Ductility of bulk nanocrystalline composites and metallic glasses at room temperature. Appl Phys Lett. 2000;77:46.

92. Sordelet DJ, Rozhkova E, Huang P, Wheelock PB, Besser MF, Kramer MJ, Calvo-Dahlborg M, Dahlborg U. Synthesis of Cu47Ti34Zr11Ni8 bulk metallic glass by warm extrusion of gas atomized powders. J Mater Res. 2002;17:186.
93. Bae DH, Lee MH, Kim DH, Sordelet DJ. Plasticity in Ni59Zr20Ti16Si2Sn3 metallic glass matrix composites containing brass fibers synthesized by warm extrusion of powders. Appl Phys Lett. 2003;83:2312.
94. Schroers J, Nguyen T, Croopnick GA. A novel metallic glass composite synthesis method. Scripta Mater. 2007;56:177.
95. Schroers J, Samwer K, Szuecs F, Johnson WL. Characterization of the interface between the bulk glass forming alloy Zr41Ti14Cu12Ni10Be23 with pure metals and ceramics. J Mater Res. 2000;15:1617.
96. Ma H, Xu J, Ma E. Mg-based bulk metallic glass composites with plasticity and high strength. Appl Phys Lett. 2003;83:2793.
97. Launey ME, Hofmann DC, Suh JY, Kozachkov H, Johnson WL, Ritchie RO. Fracture toughness and crack-resistance curve behavior in metallic glass-matrix composites. Appl Phys Lett. 2009;94:241910-1.
98. He G, Eckert J, Loser W, Schultz L. Novel Ti-base nanostructure-dendrite composite with enhanced plasticity. Nat Mater. 2003;2:33.
99. He G, Eckert J, Loser W, Hagiwara M. Composition dependence of the microstructure and the mechanical properties of nano/ultrafine-structured Ti-Cu-Ni-Sn-Nb alloys. Acta Mater. 2004;52:3035.

Chapter 2
Fabrication Methods of Artificial Microstructures

2.1 Metallic Glass (MG) Alloy Synthesis

The exceptional processability and large supercooled liquid region (SCLR) of MGs makes them highly promising candidates for thermoplastic processing, especially when replicating features with high precision and tolerance. A lightweight $Zr_{35}Ti_{30}Cu_{7.5}Be_{27.5}$ MG former ($\rho \approx 3.9$ g cm^{-3}), which has the largest SCLR of $\Delta T = 159$ K (at 20 K/min heating rate) of any known bulk glass forming alloy is utilized for heterostructure fabrication. This alloy can be cast into fully amorphous rods of 1.5 cm in diameter, and shows enhanced glass forming ability (GFA). In addition, this alloy exhibits high-yield strength in compression ($\sigma_y = 1430$ GPa) and relatively better fracture toughness than many other MGs, and a relatively high Poisson's ratio of $v = 0.37$. The undercooled liquid exhibits an unexpectedly high Angell Fragility of $m = 65.6$ [1]. Microreplication methods carried out in open air using relatively low applied pressures (~1 atm) demonstrate superior thermoplastic processability of Zr-based alloys for engineering applications [2]. Furthermore, strain rate effects on viscosity of this alloy and similar Zr-based MG alloys have been extensively studied [1, 3], and based on these measurements, it is demonstrated that $Zr_{35}Ti_{30}Cu_{7.5}Be_{27.5}$ exhibits exceptional properties for thermoplastic processing.

To cast, the weight percent's for the constituent elements were calculated, and each element was separately measured three times with a calibrated balance. The constituents were subsequently mixed in a quartz tube in a Ti-gettered argon atmosphere under vacuum, and melted with purity higher than 99.99 % in an arc melter, which was subsequently cooled down in air. This process forms a homogenous crystalline ingot. In order to prepare amorphous samples, the cast ingot was sealed in a quartz tube under vacuum of 10^{-6} mbar, and the quartz tube was subsequently purged with an inert gas several times to eliminate any residual oxygen content. The ingot was then heated above its liquidus temperature for 5 min, and quenched in water respectively, which results in a rod-shaped MG [2].

In addition, $Pt_{57.5}Cu_{14.7}Ni_{5.3}P_{22.5}$ MG with a low TPF processing temperature (~250–280 °C) and outstanding formability was used [4]. Its processing temperature is lower than the melting point of most metals, and has very high corrosion

B. Sarac, *Microstructure-Property Optimization in Metallic Glasses*, Springer Theses,
DOI 10.1007/978-3-319-13033-0_2

and wear resistance. The glass transition and crystallization temperatures for the $Pt_{57.5}Cu_{14.7}Ni_{5.3}P_{22.5}$ MG are 230 and 310 °C, respectively, indicating a large processing window [5]. Extensive plasticity was observed during bending and unconfined uniaxial compression, and is a consequence of the very large Poisson ratio ($v=0.42$) of the material [4].

Pt-MG ingot was prepared by melting the elements of purity better than 99.99 % in a vacuum-sealed quartz tube. B_2O_3 was used to flux the ingot for 15 min, which was shown to increase the GFA by reducing the alloy's oxides. An amorphous rod was obtained by casting it in a quartz tube, and by subsequent water quenching. X-ray diffraction machine (XRD-6000 Shimadzu) and differential scanning calorimeter (DSC; Perkin Elmer Diamond DSC) confirmed that both MG formers are fully amorphous [2].

2.2 Silicon Mold Fabrication

Honeycomb features were designed using AutoCAD 2011 software program. Overall, a circular area of 15 cm in diameter (150 mm Si wafer) can be covered with patterned molds of the artificial microstructures. Figure 2.1 shows the schematics of the CAD drawing of various honeycomb patterns designed. The drawing layout was transferred to a photomask maker (Heidelberg DWL-66 Laser Mask Writer), which uses direct laser beam writing to fabricate the chromium-etching mask. A

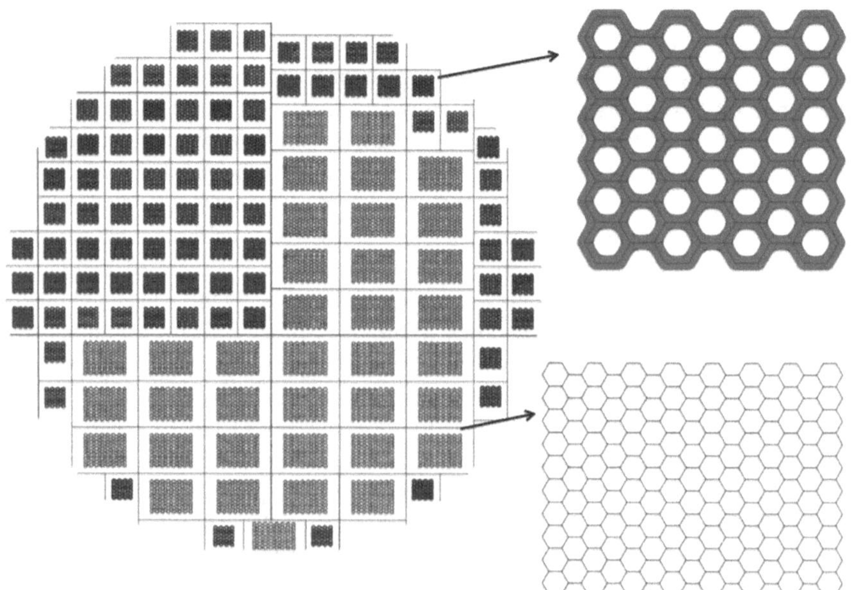

Fig. 2.1 CAD drawing of various cellular structures designed for 6″ (150 mm) wafer

Fig. 2.2 Negative Si template of a honeycomb structure fabricated by deep reactive ion etching (DRIE) process. High precision on the etched lateral walls, as well as on the edges of the features can be obtained, which is subsequently be replicated precisely by metallic glasses (MGs). Pitting is observed only on the outer surface of the mold, which is due to the overexposure of the photoresist required for high-aspect ratio etches

layer of AZ P4620 photoresist was deposited onto the silicon wafer at a speed of 3000 rpm for 42 s, which was successively patterned using basic photolithography. The final thickness of the coating was 8 μm. After the deposition of the photoresist, deep reactive ion etching (DRIE) method was utilized to each exposed areas of Si wafer. DRIE processing etch cycle was performed under 12 s, 130-sccm SF_6, 13-sccm O_2, 600-W coil power, 12-W platen power, and 26-mtorr pressure, and the passivation cycle was performed under 8 s, 85-sccm C_4F_8, 600-W coil power, 0-W platen power, and 15-mtorr pressure. The etching depth of the honeycomb molds was measured to be around 300 ± 30 μm, where the etching depth of the cellular structure molds was controlled through the etching time. The etched samples were diced using a wafer-dicing saw. The excessive resist layer was then removed from the surface by cleaning in acetone, isopropanol, and deionized water, respectively. N_2 was used to dry the Si wafer under a pressure of 2×10^5 Pa. Figure 2.2 shows an example of a honeycomb Si mold pattern with 100 μm wide feature size. It is important to note that lateral spacing of about 2 mm between the molds was kept to prevent stress concentrations between the samples during the fabrication of the Si mold.

2.3 Fabrication Methods of MG Artificial Microstructures

Novel TPF-based processing methods developed for MGs have remarkably increased the possibility of geometries that can be net shaped [6]. To exemplify, complex shapes that were previously unachievable with any metal processing method can be easily fabricated with high precision using this concept [7, 8]. A considerable attention has lately been given to the TPF of MGs on micro- and nanoscale applica-

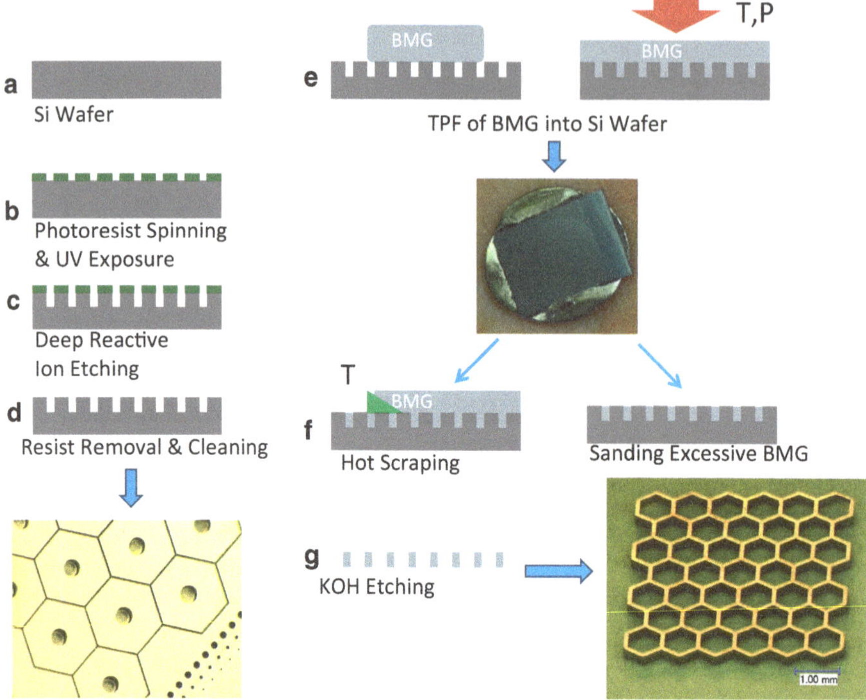

Fig. 2.3 Schematics of the MG cellular structure fabrication. **a** Silicon is used as a mold material. **b** A thin layer of photoresist on the *Si wafer* is patterned by UV exposure through a mask. **c** The exposed regions of the *Si wafer* are etched by *deep reactive ion etching* (DRIE). **d** The residual photoresist is removed. **e** The etched *Si wafer*, containing the cellular structure features, are filled with Zr- or Pt-MG by TPF based compression molding. **f** Residual MG is either removed by *hot scraping* or polishing. **g** The cellular structure of 39 cells is released by etching the *Si wafer* in KOH (Reprinted from reference [20] with a permission from John Wiley and Sons)

tions due to their homogeneous structure and ease in fabrication on multiple length scales [9–17].

We consider $Zr_{35}Ti_{30}Cu_{7.5}Be_{27.5}$ as a typical MG with negligible macroscopic plasticity [1] and, as a comparison, $Pt57.5Cu_{14.7}Ni_{5.3}P_{22.5}$,which exhibits unusually high, ~20 % compressive plasticity in their bulk forms [4]. Both alloys are well suited for the TPF fabrication of the cellular structures due to their high formability [5].

Figure 2.3 shows the fabrication steps of the MG cellular structures. Discs of 1 mm thickness cut from the amorphous MG rods were pre-pressed to a thickness of around 500 μm by using Instron 5569 tensile testing machine (50 kN maximum load capacity) via compression plates. The Si mold containing the desired pattern was heated to the TPF temperature of the related MG, and the pre-pressed MG disc was placed on the heated mold and equilibrated to attain a stable temperature. The cellular structures were produced by thermoplastic compression molding of $Zr_{35}Ti_{30}Cu_{7.5}Be_{27.5}$ at 425 and 275 °C for $Pt_{57.5}Cu_{14.7}Ni_{5.3}P_{22.5}$ in air into the etched

Si mold for 60 s under a pressure of 50 MPa. Under these conditions, the MG subsequently relaxes into a highly viscous metastable liquid, and flows into the etched mold under controlled pressure [18]. The TPF processing temperature is selected on the basis of the viscosity change as a function of temperature and processing window. A typical temperature used for TPF is chosen such that the viscosity decreases to 10^6–10^8 Pa·s as the crystallization time is approximately 3–5 min [19]. Air-cooling is sufficient to preserve its amorphous nature, which can be verified by subsequent thermal and structural analysis. The key to TPF-based compression molding is the precise temperature control, and the applied pressure should exceed the flow stress of the MG to achieve the required strain before crystallization sets in. As the viscosity of the MG in its supercooled liquid state changes by roughly an order of magnitude per every 20 °C, temperature fluctuations during net-shaping should be minimized [19].

After the pressed samples were cooled down to room temperature, the extra MG layer was removed by either hot cutting [12] at 390 °C or by polishing using Buehler Metaserv 250 Grinder-Polisher. The aspect ratio dependent etching rate of Si molds results in cellular structures varying in depth. Samples were grinded down to 200 ± 10 μm to have uniformity in depth (z-direction) of the MG cellular structures with different l/t ratio. The MG cellular structure comprised of 39 cells was subsequently released from the mold by etching the Si mold in KOH for 30 min at 100°C (Fig. 2.3g). The fully amorphous structure was confirmed by subsequent thermal and structural analysis. Similarly, MG heterostructures for tensile testing were also created using this twofold technique, where the precision of the CAD drawing can be translated into the MG within a precision of a micron (Fig 2.4).

Thermoplastic forming (TF) of MGs into Si molds causes residual stresses at the matrix-second phase interface due to the linear expansion coefficient difference between the Si pillars and the MG surrounding them. Tensile heterostructures were subsequently postannealed at 380 °C for 5 min (processing window at this temperature measured by the differential scanning calorimeter is ~20 min) to dissipate the stress concentrations caused from the fabrication step. Surface oxidation caused by the postannealing process was eliminated by polishing the surface of the sample. Using this pretest method, overall fracture strength and strain of MG heterostructures improved by ~10 %, and this increase is accounted for the linear expansion of the pores through:

$$\sigma_{xx} = E(x)\alpha(x)\Delta T \tag{2.1}$$

where E (elastic modulus of the heterostructure) is taken as 20 GPa for a porous sample, α (linear expansion coefficient) is taken as 10^{-5} 1/K [21], and the temperature difference as 400 K, which results in $\sigma_{xx} = \sigma_{yy} = 80$ MPa.

The net-shaping of thin parts of MGs is restricted by friction between the mold and the viscous MG. An ideal method to eliminate friction is blow molding, which has been demonstrated for MGs on the macroscale [7]. Using this unique shaping method, it has been already shown that the formation of precise net shaping of

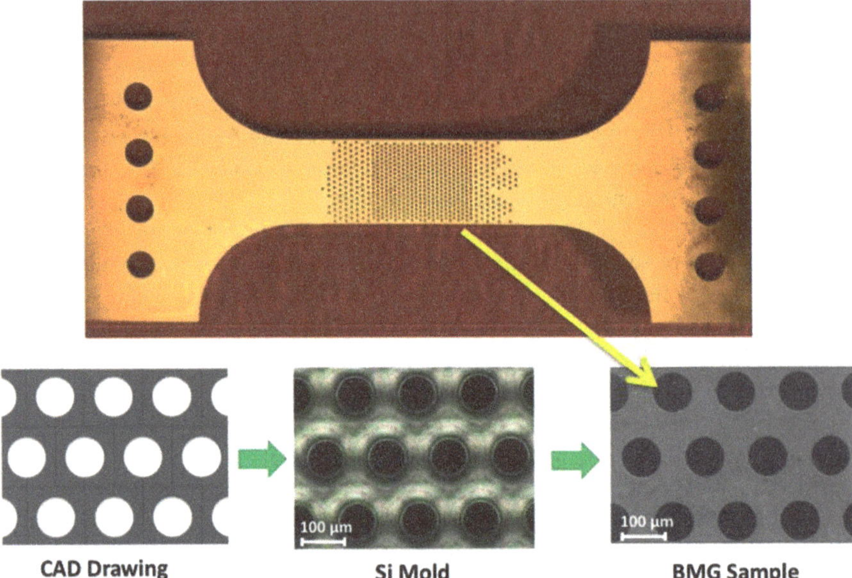

Fig. 2.4 MG heterostructure realized in a tensile test sample, and close-up image of its gauge section. Comparison of the *Si-mold* with the original *CAD drawing* reveals the control and precision of our process

complex parts over a wide range of length scales (from angstrom to meter-scale) is possible [22]. To show the experimental feasibility of the blow-molding process, our group has utilized a free-expansion technique [2]. A schematic sketch of the blow molding of 3D microshells at its processing conditions is shown in Fig. 2.5(i). The side and cross-sectional view of a blow molded MG microshell is shown in Fig. 2.5(ii-a, b). The uniformity of the shells in Fig. 2.5(ii-c) suggests that the expansion process can be well controlled on a macroscale. MGs feature the largest resistance to thinning at typical blow molding conditions, and hence, the cross-sectional profile has a very high uniformity [5]. Figure 2.5(iii) depicts the change in aspect ratio (h/d_0) during expansion as a function of temperature. A large and desired processing window is available for blow molding a hollow shape with an aspect ratio of 2 at 450 °C and above.

In addition, our group has presented a method to fabricate 3D $Pt_{57.5}Cu_{14.7}Ni_{5.3}P_{22.5}$ MG microshells attached to a Si wafer by combining TPF-based blow molding and compression molding of MG. The details of the Si mold photolithography, measures taken to protect the MG disc from flowing into the big cavities, and the TPF based compression molding are delineated in [2]. In this method, the MG-Si assembly is brought to the processing temperature of 275 °C. As a result, the unsupported MG diaphragm expands into a 3D shell while attached to the Si wafer at the same time (Fig. 2.5(iv-a, b). The height of the MG shells can be adjusted

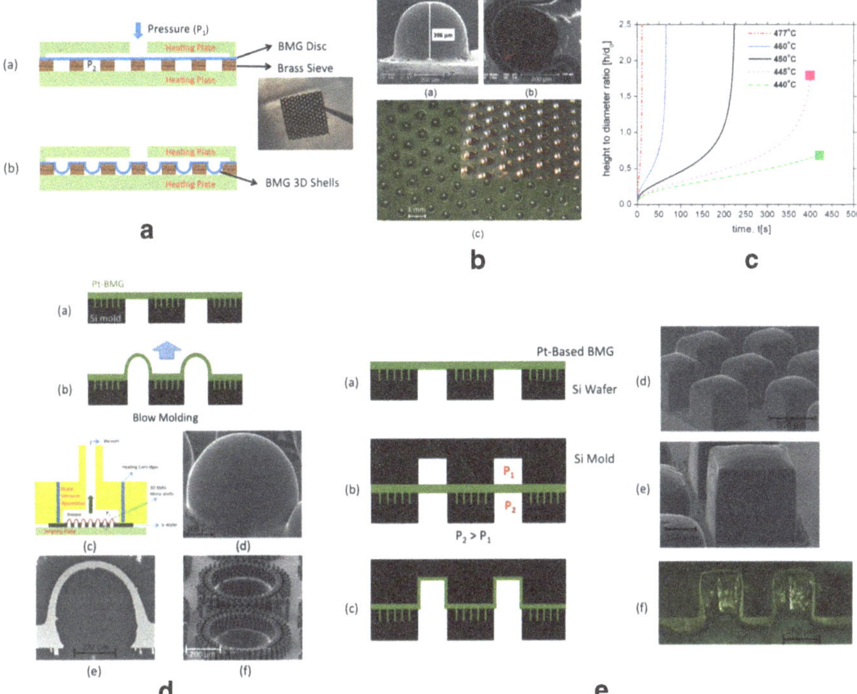

Fig. 2.5 (*i*) The $Zr_{44}Ti_{11}Ni_{10}Cu_{10}Be_{25}$ MG disc and brass sieve are gradually heated to the processing temperature of the MG of 450 °C. For blow molding, $\Delta P = 10^5$ Pa pressure difference is generated between the top and the bottom sides of the disk by applying pressure of 2×10^5 Pa to the top of the disc. (*ii*) **a** A side view and **b** 90° rotated cross-sectional view of a MG microshell taken by SEM. (**c** and inset) Array of microshells produced by blow molding through a brass sieve with 500 μm holes. (*iii*) Expansion kinetics of a $Zr_{44}Ti_{11}Cu_{10}Ni_{10}Be_{25}$ shell at various temperatures under a pressure difference of 5×10^5 Pa with a template feature diameter and initial disc thickness of 500 and 200 μm, respectively. The onset of crystallization is indicated by the solid squares which terminates the expansion. Change in the height of the 3D shell with respect to blowing time is used to estimate the expansion kinetics. (*iv*) **a-b** The pressed Pt-MG layer is blow molded through large holes where the filled small holes anchor the Pt-based MG to the Si wafer. **c** The vacuum apparatus applies a pressure gradient ($P_1 < P_2$) of 0.1 atm between the front and the backside of the MG-Si assembly. **d** A side view of a ~500 μm microshell, **e** a cross-sectional view, and **f** the MG side that is originally attached to the Si wafer is depicted. Pt57.5Cu14.7Ni5.3P22.5 is blow molded into a square mold at 275 °C for 30 s at a pressure difference of 5 t 105 Pa

by optimizing temperature, time, and pressure (see Fig. 2.5(iii)). Uniform stress distribution inside the diaphragm ensures the sphericity of MGs shells, which mainly depends on the thickness of the initial MG layer. A schematic sketch of the vacuum setup is presented in Fig. 2.5(iv-c), which was used to blow mold 3D microshells. Si seals the groove of the vacuum apparatus inserted on top of the MG-Si assembly. Figure 2.5(iv-d) shows the freely expanded Pt-based MG shell by blow molding. The Si wafer was subsequently etched using KOH to visualize the attachment of the microshells to Si via the small anchoring holes. Figure 2.5(iv-e) is the cross-sectional view of the expanded shell, where the microshell has ~40 μm thickness at the pole. SEM image of the MG side is illustrated in Fig. 2.5(iv-f), which was originally attached to the Si wafer. A strong attachment of the MG to the Si wafer is provided by the DRIE process, which produces the vertical surface roughness (scalloping effect) of ~400 nm.

The 3D shell fabrication method can also generate shells beyond the free expanded spherical shapes by blowing into mold cavities ([2]; Fig. 2.5(v)). In this technique, a second Si wafer with square cavities (500 μm each side) is used as a template. $Pt_{57.5}Cu_{14.7}Ni_{5.3}P_{22.5}$ is attached to a Si wafer using similar conditions in Fig. 2.5(iv). A second Si mold with the square cavities of the same dimensions is clamped to the MG-Si assembly and aligned with reference to the larger cavities in the original Si wafer (Fig. 2.5(v-b)). Argon gas is utilized to create a pressure difference between two sides under inert environment, which blows the attached MG into the square mold cavities (Fig. 2.5c). Fig. 2.5(v-d, e) show examples of square MG shells anchored to Si wafer. The pole section of the blow-molded shells are ~15 μm (Fig. 2.5(v-f)). This novel production technique corroborates that MG shells with controlled shapes can be created with high precision and throughput.

Conclusions

In this chapter, TPF was introduced as a net-shape processing method for MGs, which decouples fast cooling and forming processes. TPF is a highly robust and versatile shaping method, and it is very similar to the techniques used for processing thermoplastics. Table 2.1 summarizes the wide range of TPF methods compared to conventional casting [22]. Due to the versatility and high precision, TPF process is ideal for replicating small features and thin sections with high aspect ratios, which makes this process appropriate for microelectromechanical systems (microfluidic devices, microthrusters), nano- and microtechnology (resonators, acoustic transducers, packaging, microlenses, and microvalves), jewelry, biomedical (superelastic springs, stents, neural impulse actuators) and optical applications, and data storage. Furthermore, highly spherical 3D shells created by blow molding with controlled roughness of less than 2 nm grants MGs an inherently high Q-factor for resonator applications, as well as for microelectromechanical systems (MEMS) or complementary metal-oxide semiconductor (CMOS) devices [2].

Table 2.1 Process-selection map for generic shapes that can be fabricated with various TPF-based MG fabrication methods (Reprinted from [22] reference with a permission from John Wiley and Sons)

Process	Process based on	Shape/ Size	Shape/ Aspect ratio	Shape/ Feature	Process level of maturity	BMC former used	Surface finish	Dimensional accuracy	Issues/Comment
Compression molding; Injection molding	TPF	Millimeters to centimeters	1–10	Undercuts open	R&D medium	$Zr_{44}Ti_{11}Cu_{10}Ni_{10}Be_{25}$ [81,82] $Pd_{40}Ni_{10}Cu_{30}P_{20}$ [117] $Zr_{35}Ti_{30}Be_{27.5}Cu_{7.5}$ [292]	Very good	Good	Mold release; mold wear due to high required forming pressures can be combined with surface patterning <1 μm
Blow molding	TPF	10 μm to centimeters (10+)	10–1000	Hollow, undercuts; low symmetry	R&D high	$Zr_{44}Ti_{11}Cu_{10}Ni_{10}Be_{25}$ [286] $Pt_{57.5}Cu_{14.7}Ni_{5.3}P_{22.5}$ [291] $Au_{49}Ag_{5.5}Pd_{2.3}Cu_{26.9}Si_{16.3}$ [291] $Zr_{35}Ti_{30}Be_{27.5}Cu_{7\cdot5}$ [291] $Zr_{53}Ti_{5}Cu_{20}Ni10Al_{12}$ [240] $Zr_{65}Al_{10}Ni_{10}Cu_{15}$ [364]	Excellent	Excellent	Most versatile shapes possible. Can be combined with joining and surface patterning < 10 μm. Size is limited by the maximum casting thickness of preshape only
Hot-rolling	TPF	Millimeters to centimeters (10+)	100+	Sheets	R&D low	$Zr_{44}Ti_{11}Cu_{10}Ni_{10}Be_{25}$ [285]	Good	Moderately good	Temperature-control critical; aspect ratio and shape size can be increased to ∞ in a continuous process

Table 2.1 (continued)

Process	Process based on	Shape/ Size	Shape/ Aspect ratio	Shape/ Feature	Process level of maturity	BMC former used	Surface finish	Dimensional accuracy	Issues/Comment
Extrusion	TPF	Millimeters to meters	100	Long fixed cross-sectional profile, hollow	R&D low	CuTiZrNi [326,327,329] $Mg_{85}Y_{10}Cu_5$ [410] $Zr_{41.2}Ti_{13.8}Cu_{12.5}Ni_{10}Be_{22.5}$ [411] $Zr_{65}Al_{10}Ni_{10}Cu_{15}$ [323] $Ti_{50}Cu_{18}Ni_{22}Al_4Sn_6$ [412] $Mg_{5.8}Cu_{31}Y_6$ [413] $Zr_{44}Ti_{11}Cu_{10}Ni_{10}Be_{25}$ [333]	Good	Moderate	Swelling ~ 15% has to be considered; Sharp corners round off
Micro-imprinting (miniature imprinting, hot-embossing)	TPF	Micrometers to millimeters	1–10	Surface patterning	R&D industrialization	$La_{55}Al_{25}Ni_{20}$ [414] $Pt_{48.75}Pd_{9.75}Cu_{19.5}P_{22}$ [298,415] Zr-Al-Cu-Ni [296,416,417] $Zr_{44}Ti_{11}Cu_{10}Ni_{10}Be_{25}$ [245,418,419] $Pt_{57.5}Cu_{14.7}Ni_{5.3}P_{22.5}$ [245] Mg-Cu-Y [313] $Au_{49}Ag_{5.5}Pd_{2.3}Cu_{26.9}Si_{16.3}$ [245] $Zr_{41.2}Ti_{13.8}Cu_{12.5}Ni_{10}Be_{22.5}$ [302] $Zr_{46.8}Ti_{8.2}Cu_{7.5}Ni_{10}Be_{27.5}$ [420] $Pd_{40}Ni_{40}P_{20}$ [303,418] $Ce_{68}Al_{10}Cu_{20}Nb_2$ [50]	Very good	Good	Typically disposable molds; can also be combined with compression molding, hot-rolling, injection molding, and blow molding
Miniature manipulation	TPF	Nanometers to millimeters	~ 1		R&D low	$Pd_{76}Cu_7Si_{17}$ [289,421] $Zr_{44}Ti_{11}Cu_{10}Ni_{10}Be_{25}$ [245] $Pt_{57.5}Cu_{14.7}Ni_{5.3}P_{22.5}$ [245,295,422] $Au_{49}Ag_{5.5}Pd_{2.3}Cu_{26.9}Si_{16.3}$ [245]			Bending, scrapping, smoothening

Table 2.1 (continued)

Process	Process based on	Shape/ Size	Shape/ Aspect ratio	Shape/ Feature	Process level of maturity	BMC former used	Surface finish	Dimensional accuracy	Issues/Comment
Miniature fabrication	TPF	Micrometers to millimeters	1–10	JD parts; No undercuts; Design freedom in 2D, other dimension perpendicular	R&D high	$Pt_{57.5}Cu_{14\cdot7}Ni_{5.3}P_{22.5}$ [245] $Au_{49}Ag_{5.5}Pd_{2.3}cu_{26.9}Si_{16.3}$ [245] $Zr_{44}Ti_{11}Cu_{10}Ni_{10}Be_{25}$ [245]	Very good to excellent	Good	Combines microimprinting, scrapping and/or manipulation, and surface smoothening
Nano-imprinting	TPF	<100 nm	1–100	Rods	R&D medium	ZrAICuNi [290] $Pt_{57.5}CU_{14\cdot7}Ni_{5.3}P_{22.5}$ [280] $Au_{49}Ag_{5.5}Pd_{2.3}Cu_{26.9}Si_{16.3}$ [280]	Very good	Good	Relies on BMG-mold material combinations
Die-casting	Direct casting	Millimeters to 10 centimeter	1–100	Undercuts	Commercially used	ZrAICuNi [235] $Zr_{41.2}Ti_{13.8}Cu_{12.5}Ni_{10}Be_{22.5}$ [82] $Zr_{57}Nb_{5}Cu_{15.4}Ni_{12.6}Al_{10}$ [82]	Moderate	Moderately good	Internal stresses, casting defects, porosity; Sliders needed for undercuts
Counter gravity casting	Direct casting	Millimeter to 10 centimeter	1–100	Undercuts	R&D high		Good	Good	Significant reduction of casting defects over die-cast articles; Sliders needed for undercuts

References

1. Duan G, Wiest A, Lind ML, Li J, Rhim WK, Johnson WL. Bulk metallic glass with benchmark thermoplastic processability. Adv Mater. 2007;19:4272.
2. Sarac B, Kumar G, Hodges T, Ding SY, Desai A, Schroers J. Three-dimensional shell fabrication using blow molding of bulk metallic glass. J Microelectromech S. 2011;20:28.
3. Lu J, Ravichandran G, Johnson WL. Deformation behavior of the Zr41.2Ti13.8CU12.5Ni10Be22.5 bulk metallic glass over a wide range of strain-rates and temperatures. Acta Mater. 2003;51:3429.
4. Sarac B, Ketkaew J, Popnoe DO, Schroers J. Honeycomb structures of bulk metallic glasses. Adv Funct Mater. 2012;22:3161.
5. Schroers J, Johnson WL. Ductile bulk metallic glass. Phys Rev Lett. 2004;93:255506-1.
6. Schroers J. On the formability of bulk metallic glass in its supercooled liquid state. Acta Mater. 2008;56:471.
7. Schroers J, Paton N. Amorphous metal alloys form like plastics. Adv Mater Process. 2006;164:61.
8. Schroers J, Pham Q, Peker A, Paton N, Curtis RV. Blow molding of bulk metallic glass. Scripta Mater. 2007;57:341.
9. Schroers J, Hodges TM, Kumar G, Raman H, Barnes AJ, Quoc P, Waniuk TA. Thermoplastic blow molding of metals. Mater Today. 2011;14:14.
10. Saotome Y, Itoh K, Zhang T, Inoue A. Superplastic nanoforming of Pd-based amorphous alloy. Scripta Mater. 2001;44:1541.
11. Sharma P, Kaushik N, Kimura H, Saotome Y, Inoue A. Nano-fabrication with metallic glass - an exotic material for nano-electromechanical systems. Nanotechnology. 2007;18:035302-1.
12. Kumar G, Tang HX, Schroers J. Nanomoulding with amorphous metals. Nature. 2009;457:868.
13. Schroers J, Pham Q, Desai A. Thermoplastic forming of bulk metallic glass—a technology for MEMS and microstructure fabrication. J Microelectromech S. 2007;16:240.
14. Chu JP, Wijaya H, Wu CW, Tsai TR, Wei CS, Nieh TG, Wadsworth J. Nanoimprint of gratings on a bulk metallic glass. Appl Phys Lett. 2007;90:034101-1.
15. Kumar G, Schroers J. Write and erase mechanisms for bulk metallic glass. Appl Phys Lett. 2008;92:031901-1.
16. Pan CT, Wu TT, Chang YC, Huang JC. Experiment and simulation of hot embossing of a bulk metallic glass with low pressure and temperature. J Micromech Microeng. 2008;18:025010-1.
17. Inoue A, Nishiyama N. New bulk metallic glasses for applications as magnetic-sensing, chemical, and structural materials. Mrs Bull. 2007;32:651.
18. Henann DL, Srivastava V, Taylor HK, Hale MR, Hardt DE, Anand L. Metallic glasses: viable tool materials for the production of surface microstructures in amorphous polymers by micro-hot-embossing. J Micromech Microeng. 2009;19:115030-1.
19. Schroers J. The superplastic forming of bulk metallic glasses. Jom-Us. 2005;57:35.
20. Kumar G, Desai A, Schroers J. Bulk metallic glass: the smaller the better. Adv Mater. 2011;23:461.
21. Ohsaka K, Chung SK, Rhim WK, Peker A, Scruggs D, Johnson WL. Specific volumes of the Zr41.2Ti13.8Cu12.5Ni10.0Be22.5 alloy in the liquid, glass, and crystalline states. Appl Phys Lett. 1997;70:726.
22. Schroers J. Processing of bulk metallic glass. Adv Mater. 2010;22:1566.

Chapter 3
Structural Characterization of Metallic Glasses

The amorphous nature and the forming characteristics of the cast metallic glass (MG) rod need to be closely monitored to assess the suitability of the cast material for the artificial microstructures. For this reason, three different test methods, namely, formability test, structural and thermal analysis, and bending test were conducted prior to fabrication and testing of MG heterostructures, and the results are presented in this chapter.

3.1 Formability Test

The key material property that determines an MG's suitability for thermoplastic forming (TPF) is the formability, where Schroers [1] have recently suggested a standard to characterize the formability of MGs, which is simple, precise, and can be carried out on most MG-forming alloys. This method measures the deformation of a specified volume of the MG heated through the supercooled liquid region under a constant load, and the maximum diameter of the disc is taken as a measure of the MG's formability.

A disc with dimensions 0.1 cm^3cut from the $Zr_{35}Ti_{30}Cu_{7.5}Be_{27.5}$ MG rod is thermoplastically deformed during heating through supercooled liquid region at a constant rate of 20 K min^{-1} as the MG disc is forced between the plates of the Instron machine with a constant load of 4500 N (Fig. 3.1). The plates have negligible effect on the final disc diameter as long as the roughness is small compared to the final thickness of the disc. This is confirmed by comparing platens polished to a mirror finish with the ones grinded by 320 grit sandpaper. However, additional lubrication has an adverse effect on the final diameter, which is accounted for the change from stick to stick–slip condition [1].

The evaluation of the formability is quantified in the final diameter of the deformed disc after passing the supercooled liquid region (SCLR) of the MG. For the repeatability of the results, standard deviation of the disc diameter is calculated from three samples. The disc diameters are found to be around 27 mm (± 0.5 mm),

B. Sarac, *Microstructure-Property Optimization in Metallic Glasses*, Springer Theses,
DOI 10.1007/978-3-319-13033-0_3

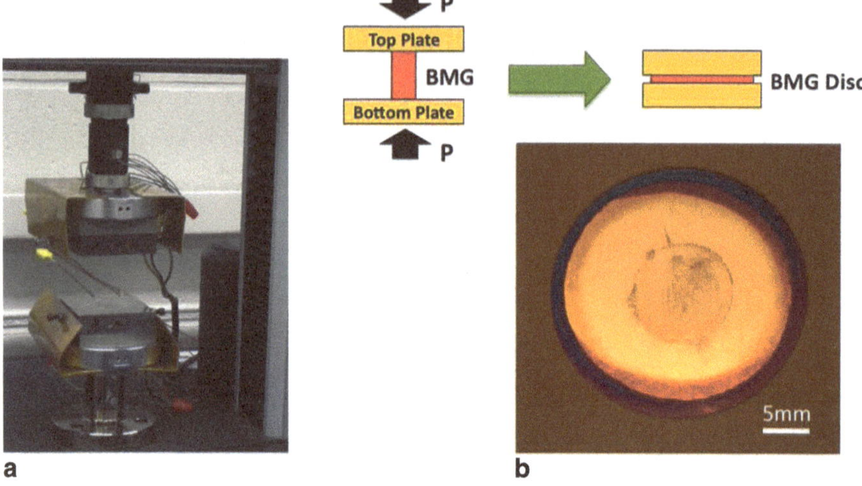

Fig. 3.1 **a** Compression molding apparatus. The metallic glass (MG) disc is squeezed between the top and bottom plates under constant applied pressure and gradual temperature ramp. **b** The final diameter of the $Zr_{35}Ti_{30}Cu_{7.5}Be_{27.5}$ MG disc is ~27 mm, which is relatively larger than the other Zr-MGs of similar composition

which is slightly above the findings for MG's of similar compositions [1], indicating a higher formability of the $Zr_{35}Ti_{30}Cu_{7.5}Be_{27.5}$ MG alloy used throughout this thesis. In addition, the resistance to oxidation of $Zr_{35}Ti_{30}Cu_{7.5}Be_{27.5}$ MG is higher than the other Zr-MGs during the formability test, which can also be noticed by the naked eye from the thermoplastically pressed discs. This is the primary reason why $Zr_{35}Ti_{30}Cu_{7.5}Be_{27.5}$ MG can be deformed to a higher extent: Surface oxidation prevents further flow and negates the attractive properties of the MG forming alloy in its amorphous state [1].

3.2 Thermal Analysis

Thermal analysis is a powerful tool to characterize MG, by which one can judge their properties as they change with temperature. This technique comprises any analysis of materials relating to heat such as freezing/boiling temperatures, heat of fusion, heat of vaporization, specific heat, etc.

The driving force for crystallization is approximated by the Gibbs free energy difference (ΔG), which is in turn correlated with the heat of fusion (ΔH_m) and the difference in specific heat capacity $\Delta c_p(T)$ between supercooled liquid and the crystal. The alloys with good glass forming ability (GFA) (which is also correlated with the formability test conducted for our Zr-MG) show shallow specific heat capacity and a strong glass forming ability, where different types of Zr-MGs (Vitreloy 1, Vitreloy 4 as well as $Zr_{35}Ti_{30}Cu_{7.5}Be_{27.5}$ MG) constitute this group [2]. The Gibbs free energy of the supercooled liquid with respect to the crystal $\Delta G(T)$ can be calculated by integrating c_p and taking ΔH_m into account [2, 3], where the lower critical cooling rate alloys (i.e., Zr-MG alloys) have lower driving force for crystallization.

Differential scanning calorimeter (DSC) is used to measure the change in the heat capacity c_p as a function of temperature and time in a controlled atmosphere. This device is able to provide quantitative and qualitative information about the physical and chemical changes that involve the amount of heat absorbed or released during endothermic and exothermic phase transitions, which is reflected in total enthalpy change (ΔH) as a function of temperature [4]. In DSC sample holder, there are two slots one of which contains an inert material such as alumina, or just an empty aluminum reference pan, and the other one contains the sample being examined. The temperature program for DSC analysis is designed such that the sample holder temperature increases linearly as a function of time. In order to keep the reference and sample pans at the same temperature throughout the experiment, the system supplies equivalent heat to the sample pan to keep the temperature of this pan increasing at the same rate as the reference pan. Figure 3.2 shows the DSC machine used in our experiments, as well as a typical thermogram of a material showing glass transition and crystallization [5].

A plot of heat capacity as a function of temperature for this type of MG is a good starting point to discuss the thermodynamics of these materials. To understand whether TPF-based compression molding of MG heterostructures has any effect on T_g, T_x, and total enthalpy of crystallization (ΔH_x), as-cast and the thermoplastically formed pieces (at 420°C for 1 min) of ~20 mg $Zr_{35}Ti_{30}Cu_{7.5}Be_{27.5}$ MG were heated through its supercooled liquid region. The DSC scans were performed at a constant heating rate of 20 K/min by Perkin Elmer Diamond DSC. Figure 3.3 compares the thermograms in the standard state and after TPF. Both samples show T_g at 314°C and the onset of crystallization in the range of 469–474°C. The onset of crystallization shifts to lower values after TPF, where there is a marginal decrease measured in the heat of crystallization (ΔH) by ~5 %. This finding suggests that the utilized fabrication process slightly affects the internal properties of the as-cast MG former, where the results are aligned with [6]. The thermal stability of the alloy remains unaltered because T_g of the alloy remains constant, and thereby, the width of the supercooled region (ΔT) shifts only by several degrees after thermoplastic forming.

3.3 Structural Analysis

Although the DSC results indicate that TPF has a subtle effect on the crystallization temperature, direct structural characterization is crucial for a comprehensive understanding of this phenomenon. X-ray diffraction (XRD) is a nondestructive technique that reveals detailed information about the crystallographic structure and chemical composition of materials [7]. In an X-ray diffraction measurement, a crystal is mounted on a goniometer, and gradually rotated while being bombarded with X-rays, where these waves undergo a phenomenon called diffraction when

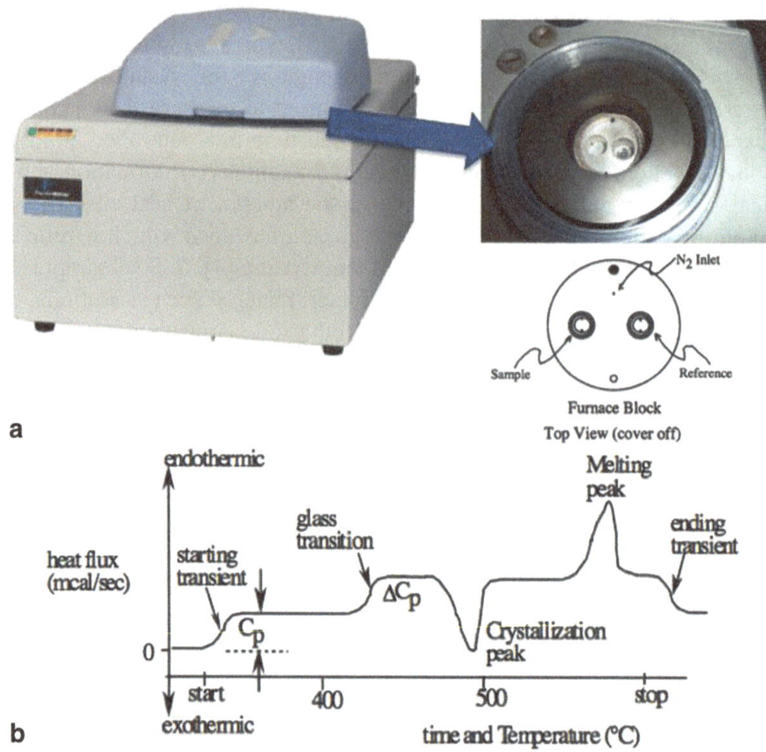

Fig. 3.2 **a** PerkinElmer Diamond differential scanning calorimeter (DSC) used in our facility, and the sample holder section for sample and reference pans. **b** Typical DSC curve illustrating the important exo- and endothermic changes

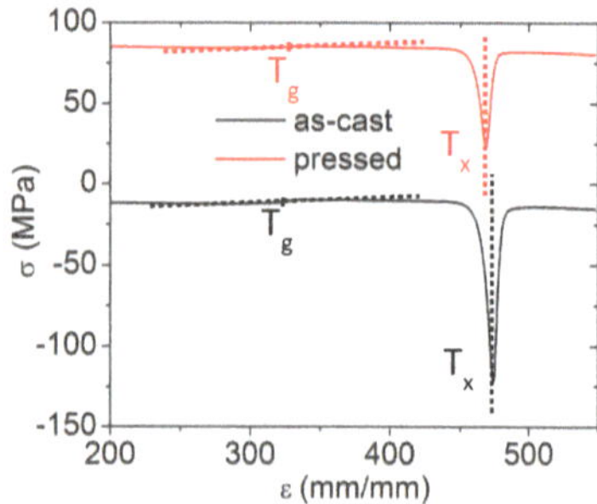

Fig. 3.3 Differential scanning calorimeter (DSC) thermograms of the as-cast and the thermoplastically formed Zr-MG show similar T_g and subtle differences in T_X and ΔH

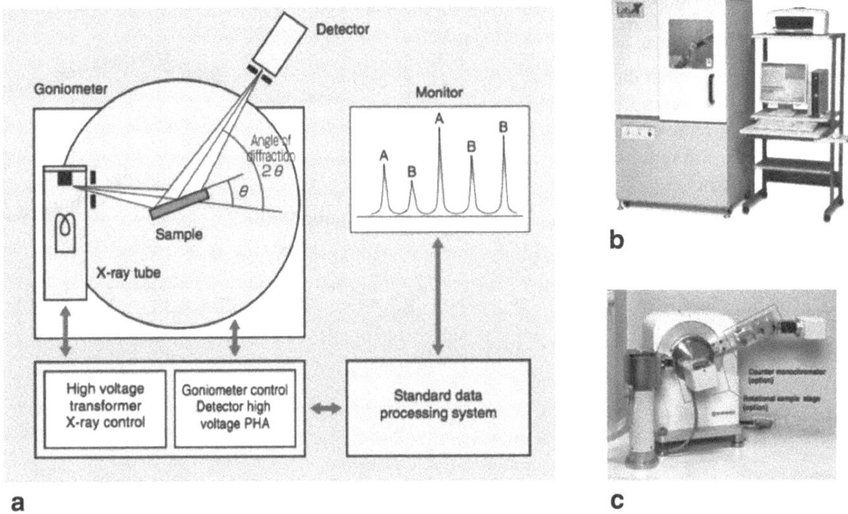

Fig. 3.4 **a** Illustration of X-ray diffraction and data acquisition/processing **b** Shimadzu X-ray diffraction (XRD) two-axis diffractometer currently used in our facility, and its vertical goniometer **c** used to obtain peaks from different angles

interacting with systems (diffracting centers), which are spaced at distances of the same order of magnitude as the wavelength of the particular radiation considered [8]. When X-rays are scattered from a crystalline solid they can constructively interfere, producing a diffracted beam. The relationship describing the angle at which a beam of X-rays of a particular wavelength diffracts from a crystalline surface is known as Bragg's Law:

$$2d \sin\theta = n\lambda \tag{3.1}$$

where λ is the wavelength of the x-ray, θ is the scattering angle, n is the integer representing the order of the diffraction peak, d is the interplane distance (i.e., of atoms, ions, molecules). If the interplane distance kept constant, these waves will be in synchronization (add constructively) only in directions where their path-length difference $2d \sin\theta$ equals to an integer multiple of the wavelength λ, and the portion of the beam is diffracted by an angle 2θ [7]. Figure 3.4a shows the schematics of the X-ray data analysis. In this study, structural analyses of the as-cast and thermoplastically formed sample were conducted by Shimadzu XRD-6000 diffractometer (Fig. 3.4b) [9]. The sample is placed on a vertical goniometer, which has two-axis rotation (Fig. 3.4c). XRD diffractograms of the as-cast and the thermoplastically formed samples show no detectable peaks (except the 38.43° peak, which is the miller indices of (100) plane for the aluminum holder the sample was placed onto), reflecting the MG's similar amorphous nature even after the fabrication of MG heterostructures (Fig. 3.5).

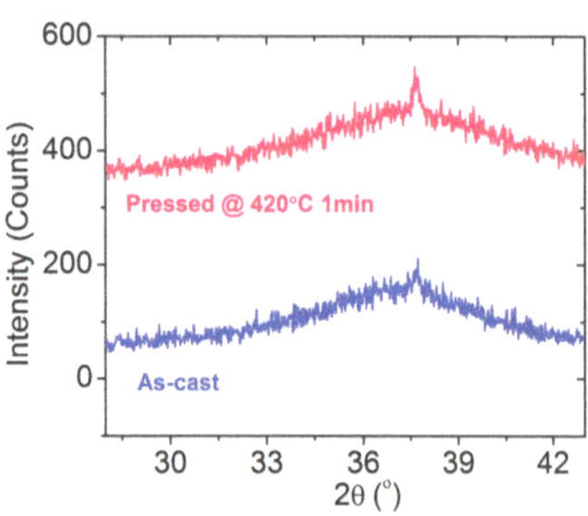

Fig. 3.5 Structural comparison using X-ray diffraction. Both samples show a broad peak, which verifies that the thermoplastic-based compression molding does not cause any measurable difference on crystallization kinetics

3.4 Bend Test

Previous methods used in this chapter provide information about thermal and structural properties of MGs, but do not reflect whether TPF creates any change in mechanical properties. Conner et al. [10] found that if one dimension of the sample goes below ~1 mm, significant bending plasticity can be observed. To measure the amount of deformation in $Zr_{35}Ti_{30}Cu_{7.5}Be_{27.5}$ MG, rectangular beams of 0.65±0.05 mm thick were prepared from the prepressed discs at 390°C for 1 min. All sides of the beams were mirror polished, and the corners were rounded to avoid stress concentration during bend test. Beams were bent over mandrels with decreasing radius from 30 mm to 5.5 mm with corresponding strains of 1.2–6.3 % under constant compressive strain rate of 10^{-3} s^{-1}. The amount of total deformation (ε_f) at each radius is calculated through $\varepsilon_f = t/2R$, with t: thickness of the sample (constant), and R: radius of the mandrel. This range spans the elastic to plastic strain region, which enables to observe the increase in the bending ductility with high accuracy. Figure 3.6 shows the setup and the beam that has been bent up to 6.3 % of strain (~4.3 % plastic deformation), respectively, where the results are in line with [10]. Beam with the same dimensions prepared from the as-cast sample was also deformed until 6.3 % total strain, with a subsequent elastic comeback of ~2.0 %. Since monolithic BMGs are known to show almost zero plasticity after reaching the elastic limit under tension, we therefore confirmed that the thermoplastic forming does not alter the properties of the as-cast MG. The shear band patterns on the bent or fractured beams were examined by scanning electron microscopy (SEM) (Fig. 3.6d) The SEM images indicate that the sample showing ~4.3 % plasticity undergoes a significant plastic deformation while forming multiple shear bands on both tensile and compression sides of the MG beam. However, shear bands on the tensile side impede when reaching the neutral axis of the bent sample due to counteracting compressive forces. Thus, it was observed that samples with thickness < 1 mm reveal extensive plasticity compared to the bending of a rectangular MG beam of > 1 mm, which breaks around 2.0 % strain and does not show any plasticity.

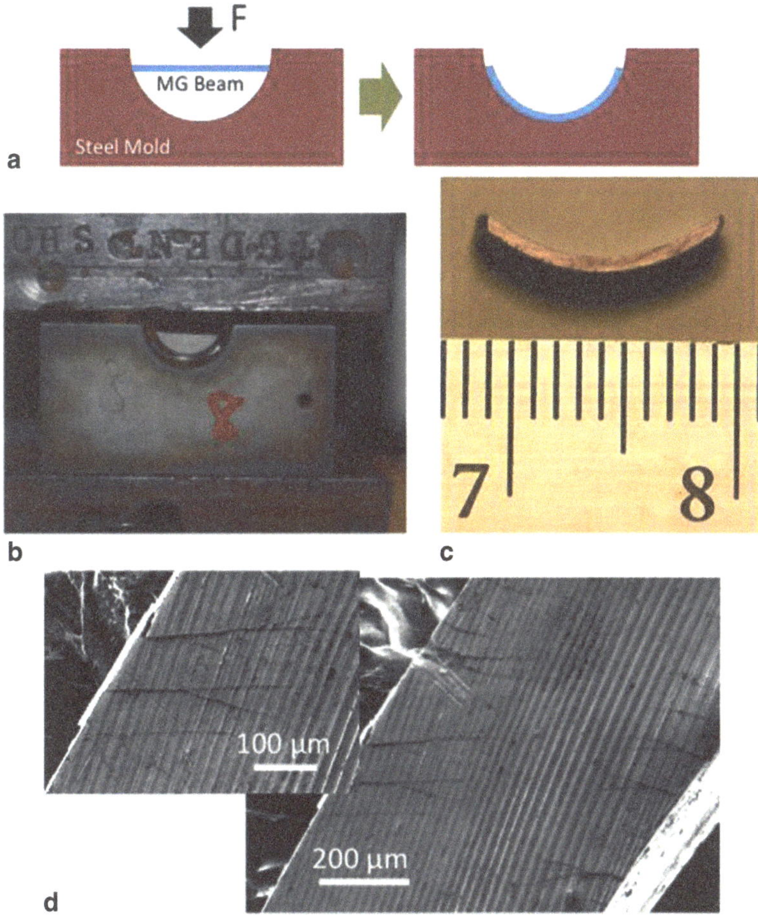

Fig. 3.6 **a** Three-point bending test, rectangular metallic glass (MG) beam bent over the mandrel. **b** Experimental setup and **c** the optical microscopy image of ~4.3 % plastically deformed specimen. **d** Scanning electron microscopy (SEM) image of the bent beam in (**c**) confirms that shear bands formed in the compression–tension zones, including shear steps formed on the tensile side (*inset*)

Conclusions

Formability test, thermal and structural analysis, and bend test are used to evaluate as-cast and TPF-based compression molded $Zr_{35}Ti_{30}Cu_{7.5}Be_{27.5}$ MG discs. The samples are showing similar mechanical, physical, and structural properties, where the results are repeatable with high accuracy. Same properties were also tested for $Zr_{35}Ti_{30}Cu_{7.5}Be_{27.5}$ MG heterostructures, where the results are exactly the same with thermoplastically deformed monolithic MGs. Therefore, it can be concluded that the properties of the fabricated structures are independent of the thermoplastic forming process and the replication of the MG heterostructures.

References

1. Schroers J. On the formability of bulk metallic glass in its supercooled liquid state. Acta Mater. 2008;56:471.
2. Busch R, Schroers J, Wang WH. Thermodynamics and kinetics of bulk metallic glass. Mrs Bull. 2007;32:620.
3. Wang WH, Dong C, Shek CH. Bulk metallic glasses. Mat Sci Eng R. 2004;44:45.
4. Differential Scanning Calorimetry. Wikipedia, The Free Encyclopedia. http://en.wikipedia.org/w/index.php?title=Differential_scanning_calorimetry&oldid=638189917. Accessed 1 Sept 2014.
5. Diamond DSC Differential Scanning Calorimeter. nthko corporation. http://www-omcs.materials.ox.ac.uk/uploads/diamond%20dsc%20tech%20sheet.pdf. Accessed 1 Sept 2014.
6. Kumar G, Rector D, Conner RD, Schroers J. Embrittlement of Zr-based bulk metallic glasses. Acta Mater. 2009;57:3572.
7. X-ray crystallography. Wikipedia, The Free Encyclopedia. http://en.wikipedia.org/w/index.php?title=X-ray_crystallography&oldid=640824627. Accessed 11 Sept 2014.
8. X-ray scattering techniques. Wikipedia, The Free Encyclopedia. http://en.wikipedia.org/w/index.php?title=X-ray_scattering_techniques&oldid=598920515. Accessed 10 March 2014.
9. LabX XRD-6000 X-ray Diffractometer. Shimadzu Corporation. http://www.ssi.shimadzu.com/products/literature/xray/xrd-6000.pdf. Accessed 10 March 2014.
10. Conner RD, Johnson WL, Paton NE, Nix WD. Shear bands and cracking of metallic glass plates in bending. J Appl Phys. 2003;94:904.

Chapter 4
Artificial Microstructure Approach

4.1 Objectives

Length scales have been identified as the main criteria in the design of effective metallic glass (MG) heterostructures [1]. Spanning six orders of magnitude (nm to mm), these length scales include different regions like shear transformation zones [1–5], where the collective shearing of the atoms in the order of a nanometer in size takes place [6]. The second region we are interested in includes the critical plastic zone size, which varies among MG formers between 10 μm and a few millimeters [7]. This is a very important length scale in the design of MG heterostructures. For example, effective MG composites exhibit a heterostructure with the second-phase spacing, which is comparable to the spacing of the plastic zone size [8–14]. It was suggested that, in this case, shear bands do not transform into cracks but shear can be absorbed into the softer second phase [3].

Such length scales have been considered widely in the design of MG composites [3, 15]. But here is the challenge: Even though length scales have been identified and experimentally confirmed, this has been done under simplified conditions. If and how such mechanisms affect mechanical behavior of real microstructures is mostly unknown. The reason for this limitation is that synthesis methods typically do not allow systematic variation individual of microstructural feature, i.e., composites/stochastic foams [2, 3, 16–23] and microstructural architecture designed by perforation stretching [24, 25] provide limited control over microstructural features, especially when independent manipulation of features is desired.

As discussed in Chap. 1, when MGs are used in geometries where one dimension is below about 10 times its critical crack length (~1 mm for a medium range metallic glass), they exhibit significant bending plasticity [26, 27]. This feature, and other size effects have been widely explored in foams to design overall plasticity [1]. In addition to the potential for microstructural architecture design, the unusual high ratio of yield strength over modulus of MGs suggests that a transition from plastic yielding to elastic buckling can be realized for practical l/t ratios (or relative densities), where l is the ligament length and t is the ligament thickness of each cell of MG cellular structure [2, 25, 28–32].

Despite high strength and elasticity combined with plastic-like processing ability, wide spread proliferation of MGs in structural applications has been generally

B. Sarac, *Microstructure-Property Optimization in Metallic Glasses*, Springer Theses,
DOI 10.1007/978-3-319-13033-0_4

stymied by their lack of plasticity [33]. Upon yielding, shear strain in MGs is highly localized in narrow, approximately 10 nm wide bands. The development of such shear bands per se does not necessarily result in a fracture. For example, when shear bands are spatially confined under certain loads and in small sizes, global plasticity enabled by the formation of a large number of shear bands has been observed [34]. In compression and bending, as well as in several experiments under tension, multiple shear band formation without crack formation has been reported [1–5, 10, 35–42]. In nanosized samples, it has been argued through a Griffith-like criterion that the elastic energy release does no longer offset the energy barrier to form a shear band [10, 11]. Hence, on such a small scale, a deformation mechanism not relying on shear bands can result in tensile ductility [40].

As an example, application for our artificial microstructure approach, we utilized two different MG heterostructures to overcome the main limitations in fabrication and characterization: Periodic cellular structures under in-plane compression to analyze the deformation behavior with respect to its relative density, and periodic porous tensile structures with controlled features to generate tensile ductility. Both topics will be introduced, and the results will be presented and discussed within this chapter.

4.2 Periodic Cellular Structures of MGs

Here, we present a method that allows varying individual microstructural features independently with unprecedented versatility and accuracy. To fabricate MG cellular structures (honeycombs), we utilize thermoplastic forming (TPF) of MGs [42, 43], and using Si molds that are produced through photolithography. Regular hexagonal cellular structures ($\theta = 30°$ and $h = l$) are used as a case study since they are widely studied. MGs are suitable to be used in such structures, and mathematical models have been developed [44–47]. To cover the range of elastic buckling to catastrophic failure, 2D cellular structures made from $Zr_{35}Ti_{30}Cu_{7.5}Be_{27.5}$ bulk MG with varying relative densities (ρ^*/ρ_s) realized in periodic hexagonal cellular structures ranging from 2.5 to 86.0 % were fabricated and characterized under quasistatic compression loading. The range of MG cellular structures was implemented to identify the different deformation modes, which are controlled by the presence of size effects as a function of density. As a result, we revealed three distinctive deformation regions: collective buckling, local failure, and global failure, and revealed the ideal density of ~25.0 % for energy absorption for the hexagonal cellular structure pattern. Furthermore, corner-fillets are used as a design tool to dissipate energy more homogenously throughout the MG cellular structures. Besides, two MGs with different characteristics are presented. Their performances in cellular structures were compared to conventional metals (superplastic aluminum alloy (2004Al), Zn) and plastics (polyether ether ketone (PEEK), polyethylene (PE)). To even extend the range of materials, we also utilized crystallization and structural relaxation of the MGs, which drastically changes their mechanical behavior [48–54].

4.2.1 MG Cellular Structure Sample

Even though unsuited for large-scale commercial manufacturing, the combination of photolithography and TPF enables a very precise, highly versatile, and high-throughput test sample fabrication. Typically, approximately 200 silicon molds can be fabricated simultaneously at a precision of better than one micron in the lateral dimensions (Fig. 4.1). This precision is translated into the MG cellular structure through TPF [15, 42], where the only measurable discrepancy between the MG cellular structures and the Si mold is due to the difference in the thermal expansion coefficients ($\alpha_{MG} \sim 4 \times 10^{-5}$/°C [55], $\alpha_{Si} = 3.7 \times 10^{-6}$/°C) [56]. To demonstrate the precision, Fig. 4.1c depicts both the feature, which was drawn in the mask, and its realization in the MG through TPF.

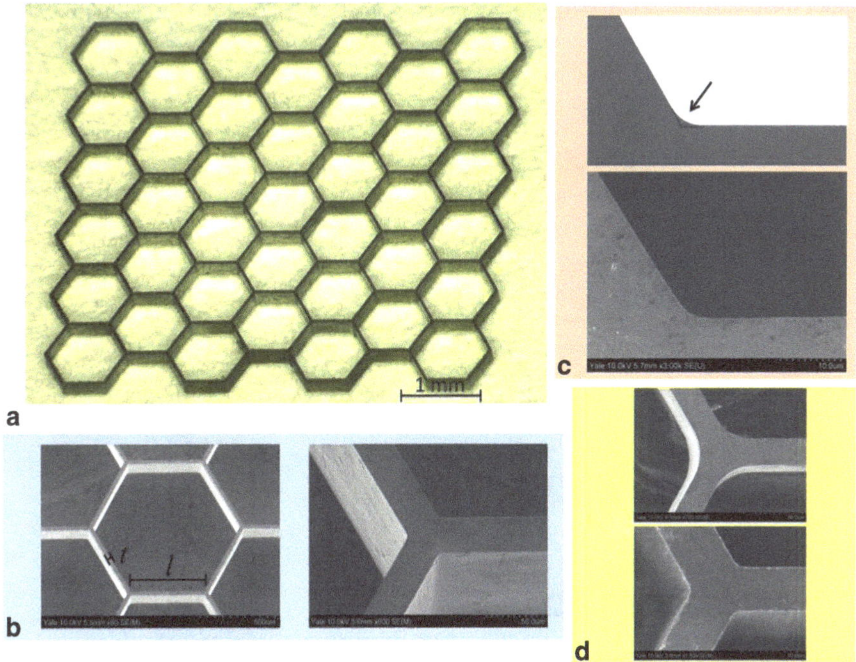

Fig. 4.1 **a** Zr-MG cellular structure with $\rho^*/\rho_s \approx 7.5\,\%$ and a depth of 200 μm. **b** Close-up image of a cellular structure cell, and its joint. **c** Corner-fillets to reduce stress concentrations in the cellular structure's joints. The *top* image shows the drawing, and the *bottom* image the realization in the actual MG cellular structure, which demonstrates our control of better than 1 μm dimensional accuracy. **d** Corner-fillets of $r = 50$ μm (*top*), $r = 6$ μm (*bottom*). (Reprinted from reference [42] with a permission from John Wiley and Sons)

4.2.2 In-Plane Compression Test

Our setup to characterize the cellular structures under quasistatic in-plane compression conditions is shown in Fig. 4.2. The tests were conducted by using Instron 5543 tensile testing machine of 1 kN maximum load capacity. To measure with the highest accuracy, load cells with a maximum load of 10, 100, and 1000 N were selected depending on the relative density of the cellular structure. Custom made pneumatic grips with grids were used to prevent slippage and guarantee precise alignment. The mold consists of two steel pieces for the top tensile grip, and a steel piece with a transparent glass panel for the bottom grip. Top and bottom parts, which are individually screwed to each other, include thin flat steel plates (shim stock) standing in between. The cellular structure was resting on the bottom plate, which was positioned between the metal mold and the transparent glass panel with a clearance exceeding the depth of the cellular structure. Alumina piece was placed under the base plate to increase the contrast of the cellular structure during imaging.

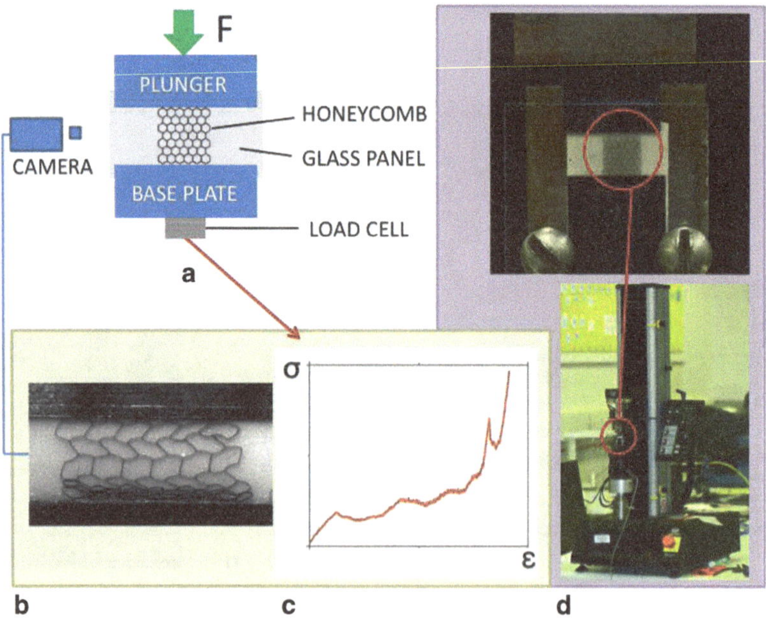

Fig. 4.2 Experimental setup for quasistatic characterization of cellular structure test samples. **a** Schematic of the mold used in this study. The cellular structure rests on a base plate, which is positioned between the metal mold and the transparent glass panel with a clearance exceeding the depth of the cellular structure. In-plane deformation of a cellular structure is achieved by pressing a flat steel plate with a constant strain rate of 0.001 s^{-1}. **b** In situ visualization of the deforming MG cellular structure. **c** Stress–strain diagram generated by the output signal of the load cell. **d** Actual experimental setup for quasistatic characterization to show the cellular structure sample positioned inside the Instron testing machine. (Reprinted from reference [42] with a permission from John Wiley and Sons)

Using Instron Bluehill Software, in-plane compression of the cellular structure was achieved by pressing the top plate with a controlled strain rate of 0.001 s^{-1} in air.

This method allowed us to simultaneously record stress–strain diagrams and visualize the deformation, hence, to correlate microstructural events with the stress–strain diagram. In situ recording through the transparent glass panel during the deformation of the cellular structure was carried out by Keyence Digital Microscope VHX-500F. Scanning electron microscopy (SEM) imaging with Hitachi SU-70 was conducted after the fracture to determine the microstructural changes, shear bands, and shear steps caused by deformation [42].

4.2.3 Euler Buckling Instability

During in-plane compression test, out-of-plane buckling is prevented by designing the structure such that the critical out-of-plane buckling stress, σ_{cr}, exceeds the yield strength, $\sigma_{y,}$ of the bulk MG. Thus,

$$\sigma_{cr} = \frac{EI\pi^2}{AL^2} > \sigma_y \tag{4.1}$$

with $I = \frac{wb^3}{12}$, I is the moment of inertia, E is the Young's modulus of the bulk MG, w is the width, b is the depth, L is the length and A is cross-sectional area of the cellular structure, σ_{cr} is the critical buckling stress, and σ_y is yield strength of the bulk MG.

While the out-of-plane buckling must be prevented through proper design of the sample for accurate experiments, in-plane buckling is explored as a design tool. In-plane elastic buckling occurs due to an Euler buckling instability for l/t values exceeding a critical ratio. This critical ratio can be correlated to the material's properties, namely E/σ_y, through [42]:

$$\left(\frac{\rho_s}{\rho^*}\right)_{crit} = \frac{\sqrt{3}}{2}\left(\frac{l}{t}\right)_{crit} = \frac{1}{2\sqrt{3}}\left(\frac{E}{\sigma_y}\right)_{bulk} \tag{4.2}$$

where ρ_s and ρ^* denote the bulk density and density of the cellular structure, respectively. Elastic buckling occurs for structures with $l/t > [l/t]_{crit}$ [57]. For example, superplastic aluminum alloy (2004Al) requires $[l/t]_{crit} \cong 82$ ($E = 74$ GPa, $\sigma_y = 300$ MPa) [58], which corresponds to a low cellular structure density of $\rho^*/\rho_s = 1.4\,\%$, preventing elastic buckling when used in practically realizable cellular structure structures. On the other hand, elastic buckling can readily be realized in plastics, such as low-density polyethylene ($E = 0.4$ GPa, $\sigma_y \cong 15$ MPa; [59]), which leads to $[l/t]_{crit} = 15.0$ ($\rho^*/\rho_s = 13.0\,\%$). For MGs, the relatively low E/σ_y ratio, e.g., $Zr_{35}Ti_{30}Cu_{7.5}Be_{27.5}$ ($E = 87$ GPa, $\sigma_y = 1430$ MPa; [60]), results in $[l/t]_{crit} = 20.4$ ($\rho^*/\rho_s = 5.7\,\%$). This density can be readily achieved in MG cellular structure structures. The significantly higher strength of MGs compared to plastics suggests that yielding and buckling of MGs occurs at a stress level, which are approximately two orders of magnitude higher than yielding and buckling of plastics [42].

4.2.4 Results and Discussion

4.2.4.1 Deformation Regions of MG Cellular Structures

To assess the deformation behavior of MG cellular structures at different cell geometries, the l/t ratio (thereby the relative density) of the hexagonal cells are varied. Figure 4.3 shows quasistatic compression of a $\rho^*/\rho_s \approx 2.5\,\%$ Zr-MG cellular structure ($l = 500$ µm, $t = 10$ µm, $l/t = 50$) at a strain rate of 0.001 s^{-1}. The stress–strain curve of the in-plane compression shows a linear-elastic region (Fig. 4.3(1)). In order to determine the nonlinear elastic deformation, additional experiments of cyclic loading to a progressively higher strain reveal 25 % elastic strain. Up to the first stress drop, the so-called stress initiation σ_I, (Fig. 4.3(2)), the deformation appears homogenously distributed throughout the structure. However, it can already be anticipated where strain localization will occur (Fig. 4.3(3)). Upon further deformation, the strength decreases

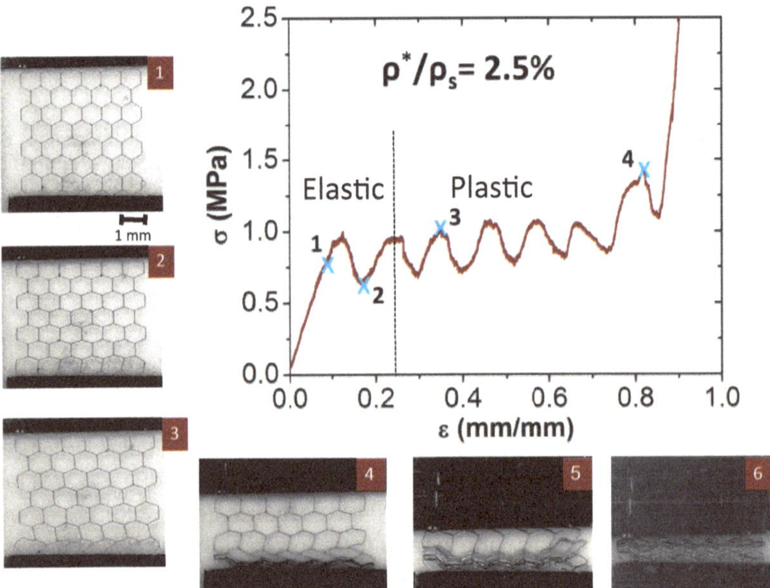

Fig. 4.3 Stress–strain diagram of in-plane compression of $Zr_{35}Ti_{30}Cu_{7.5}Be_{27.5}$ cellular structure with $\rho^*/\rho_s \approx 2.5\,\%$ ($l/t = 50$). The cellular structure exhibits in-plane elastic bending strain of 25 %, which further deforms plastically until ε_D without ligament fracture. *1–6* shows the microstructural events correlated with the stress–strain curve. (Reprinted from reference [42] with a permission from John Wiley and Sons)

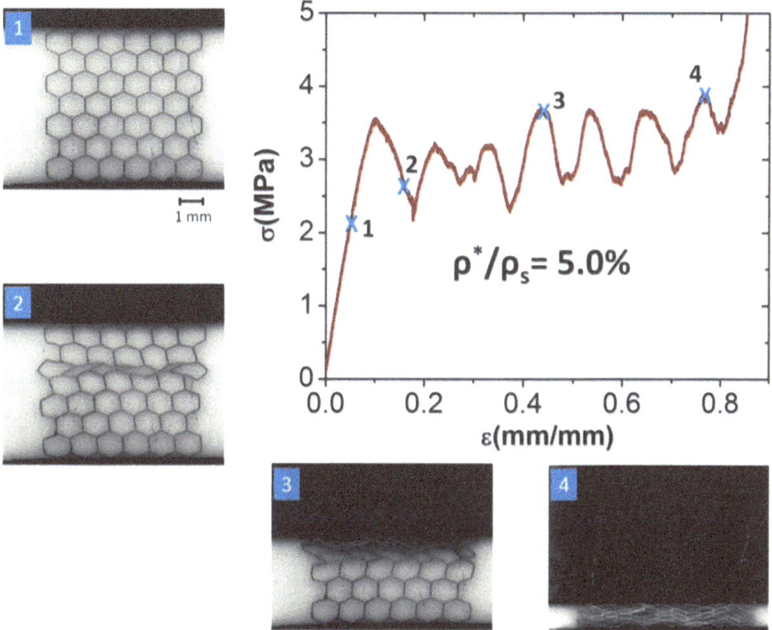

Fig. 4.4 Compressive response and fracture pattern of cellular MG structures with $\rho^*/\rho_s \approx 5.0\,\%$, showing higher stress fluctuations throughout the deformation compared to $\rho^*/\rho_s \approx 2.5\,\%$ due to higher stress levels of deformation (see Table 4.2; Reprinted from reference [15] with a permission from Elsevier)

due to the instability of the ligaments, causing the strain to localize in the bottom row of the cellular structure (Fig. 4.3(4, 5)). Such instabilities, which cause strain localization and result in stress fluctuations, are designated by seven stress hills and valleys in the stress–strain diagram. The slender cell ligaments buckle collectively and therefore exhibit a low-plateau stress, where nonlinear elastic buckling of the ligaments contributes to the high elasticity of the cellular structure of up to 12.0% [15, 42].

Similarly, the cellular structure with $\rho^*/\rho_s \approx 5.0\,\%$ ($l/t = 25$) shows collective buckling of ligaments through row-by-row deformation (Fig. 4.4). Therefore, this cellular structure exhibits a low-plateau stress, where the contribution of the nonlinear elastic component is only ~2% in this case [42].

Figure 4.5 shows the in-plane compressive deformation of $Zr_{35}Ti_{30}Cu_{7.5}Be_{27.5}$ cellular structure with $\rho^*/\rho_s \approx 24.0\,\%$ ($t = 100$ μm, $l/t = 5$). This cellular structure exhibits significantly different properties than the previous cellular structures. For $\rho^*/\rho_s \approx 24.0\,\%$, a standard linear-elastic region is observed until the initiation strength (σ_I) is reached. After exceeding the elastic limit, deformation is highly localized on the direction having the highest stress concentration (~45° angle), which is typical for the yielding of bulk MGs under compression [15, 42]. Plastic strain is highly localized within a narrow band similar to the formation of shear bands in monolithic MGs, which results in pronounced stress drops owing to fracture of the hexagonal

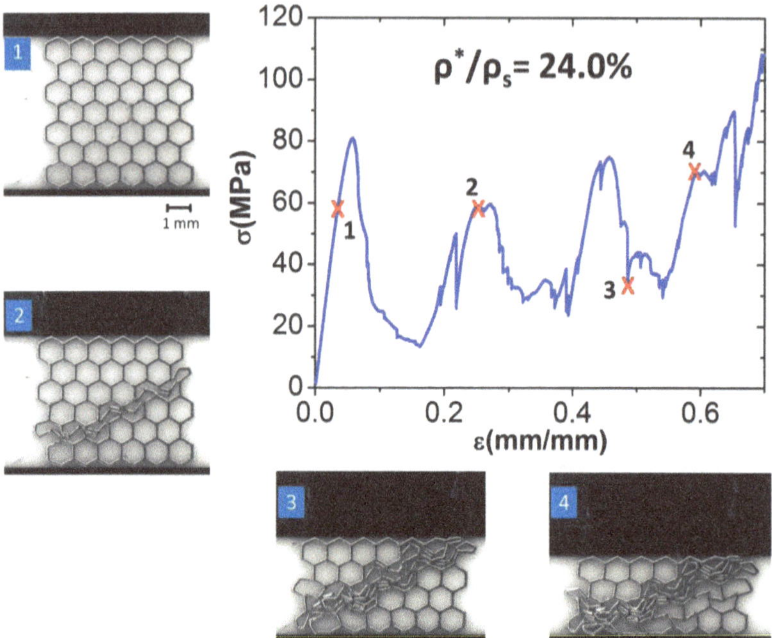

Fig. 4.5 Stress–strain diagram of in-plane compression of $Zr_{35}Ti_{30}Cu_{7.5}Be_{27.5}$ cellular structure with $\rho^*/\rho_s \approx 24.0\,\%$ ($l/t=5$). This cellular structure shows an overall elastic strain of 6 % with zero nonlinear elasticity, and beyond this point, exhibits a brittle deformation behavior with a localized deformation forming a ~45° angle with the load direction. Stress fluctuation ($\Delta\sigma_a$), which is the standard deviation of σ_P (for further definition of $\Delta\sigma_a$ see Table 4.1), increases with ρ^*/ρ_s increasing because the amount of stress to create instability at the cell ligaments increases. However, σ_a normalized by the initiation strength of the same cellular structure (σ_I) is found to be independent of the ρ^*/ρ_s of the cellular structure. (Reprinted from reference [42] with a permission from John Wiley and Sons)

cells. However, damage of the ligaments remains confined, and does not travel through the entire structure. In contrast, for the cellular structures with row-by-row deformation as shown before, yielding occurs in a plane parallel to the loading direction, which is typical for ductile materials. Not only do these two cellular structure types differ in the deformation mode, but they also show remarkable differences in their initiation stress (σ_I), and elasticity (ε_{el}) [61, 62].

In contrast, for the cellular structure with $\rho^*/\rho_s \approx 58.0\,\%$ ($t=300$ μm, $l/t=1.67$), yielding initiated by the formation of one or very few shear bands, which results in global failure of the structure (Fig. 4.6). As a consequence, no significant deformation is observed (Fig. 4.6(1, 2)), and the sample fails catastrophically with almost no plasticity (Fig. 4.6(3, 4); [15]).

Our results (Figs. 4.3, 4.4, 4.5, and 4.6) reveal different deformation behaviors in the various density ranges. Figure 4.7a depicts the various deformation modes. For $\rho^*/\rho_s < 12.0\,\%$, the collective buckling of the structure organized in row-by-row cell deformation parallel to the loading axis results in high (linear and non-linear)

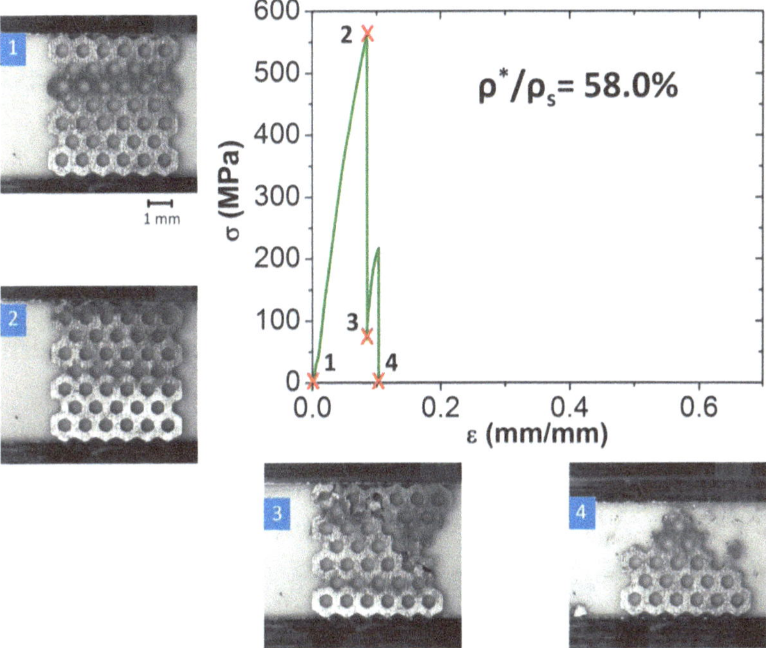

Fig. 4.6 MG cellular structure with $\rho^*/\rho_s \approx 58.0\,\%$ ($l/t = 1.67$) showing sudden fracture after exceeding σ_1. (Reprinted from reference [15] with a permission from Elsevier)

elasticity (up to 25 %) and high plasticity, however at the expense of low strength. Microscopically, this is achieved by the formation of multiple shear bands, which are occupying large parts of the ligaments (Fig. 4.7(b-i) and 4.14(iv)). The deformation mode changes to more brittle-like as the relative density increases. Localized failure is the predominant deformation mode for densities of $12.0\,\% < \rho^*/\rho_s < 40.0\,\%$. Here, shear bands transform readily into cracks (Fig. 4.7(b-ii)), however these cracks are confined into one cell. Yielding occurs prior to buckling, and shear bands transform into cracks causing local cell damage. The failure of one ligament increases stress in the vicinity and causes failure of a neighboring cell ligament at an orientation of ~45°, the plane of maximum resolved shear stress. Once an entire plane has collapsed, failure is initiated in a neighboring plane. Globally, the structure remains intact and can support load, however the strength of the structure drops significantly (down to 20 % of its yield strength).

For densities $\rho^*/\rho_s > 40.0\,\%$, the structure fails catastrophically along one or a few shear bands. The consequence is negligible global deformation prior to global fracture. This may be due to the fact that for these densities, the width of the ligaments ($t \geq 200$ μm) permits a shear band to travel throughout the entire sample, hence, once a shear band transforms into a crack, it travels along a plane of maximum resolved shear strain percolating through the ligaments [15].

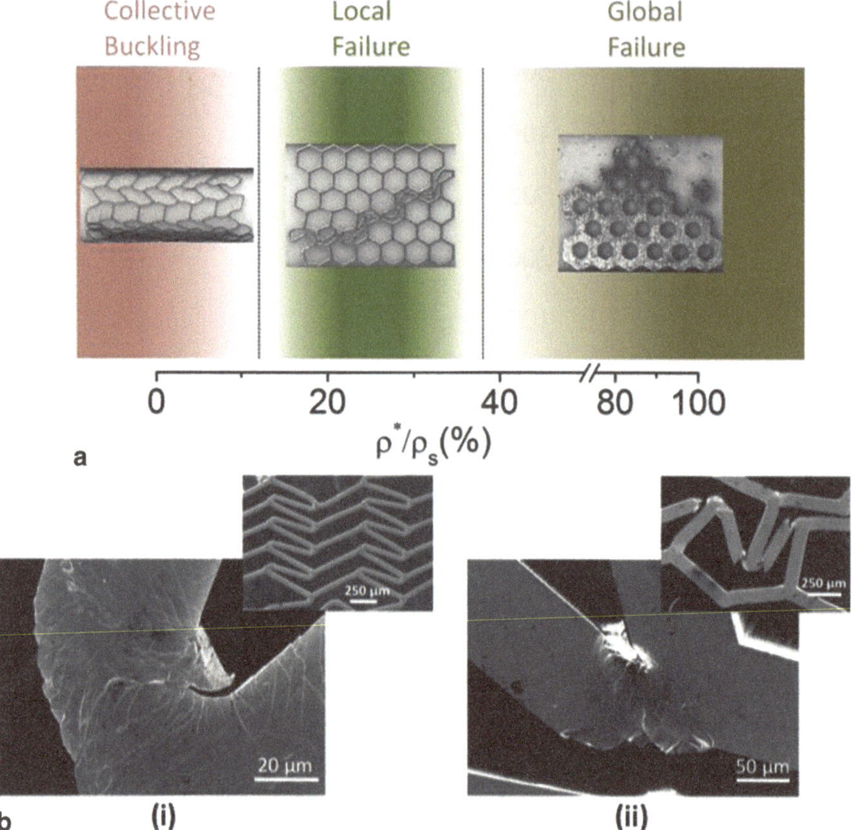

Fig. 4.7 **a** Deformation modes of cellular Zr-MG as a function of relative density. For $\rho^*/\rho_s < 12.0\,\%$, collective buckling results in high elasticity but low strength is the predominant deformation mode. Local failure of ligaments occurs for $12.0\,\% < \rho^*/\rho_s < 40.0\,\%$, whereas $\rho^*/\rho_s > 40.0\,\%$, failure percolates through the structure causing global fracture. **b** SEM images for (*i*) $\rho^*/\rho_s < 12.0\,\%$ shows multiple shear bands which are spread out over large fraction of the ligament length, where the collective buckling results in row-by-row deformation (inset). (*ii*) Shear band formation is limited to a small region at the joints for densities $18.0\,\% < \rho^*/\rho_s < 24.0\,\%$, which readily transform into cracks causing local failure of the structure (inset). (Reprinted from reference [15] with a permission from Elsevier)

4.2.4.2 Manipulation of Geometry

To explore ρ^*/ρ_s as a design tool for MG cellular structure, the ρ^*/ρ_s ratio was varied from 2.5 to 86.0 %. Figure 4.8 exhibits the stress–strain curves for $Zr_{35}Ti_{30}Cu_{7.5}Be_{27.5}$ cellular structures within this range under uniaxial in-plane compression. When increasing the density from 2.5 to 86.0 %, the initation or plateau strength increases by over three orders of magnitude from <1 to 1000 MPa at the expense of a decrease in fracture strain down to ~5 %, indicating a nonlinear relationship with density [15].

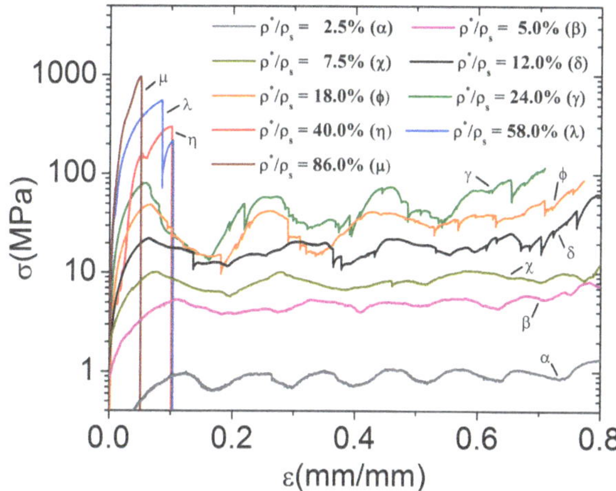

Fig. 4.8 Summary of deformation results for cellular MGs with varying relative density (ρ^*/ρ_s). Initiation strength increases by over three orders of magnitude as ρ^*/ρ_s goes from 2.5 to 86.0 %. For ρ^*/ρ_s>40.0 %, plasticity is drastically reduced due to the catastrophic failure of the cellular structure. (Reprinted from reference [15] with a permission from Elsevier)

4.2.4.3 Cellular Structures of Different Materials

To compare MG cellular structures with other materials, polyether ether ketone (PEEK) and polyethylene (PE) were chosen as representatives for plastics, and 2004Al and zinc as representatives for crystalline metals. One main criterion for the selection of these representative materials is their suitability for our fabrication process. PEEK and PE were formed below its melting temperature at 340 °C ($T_{m\,(PEEK)}$ = 343 °C) and at 110 °C ($T_{m\,(PE)}$ = 120 °C) under a pressure of 10 MPa, respectively. Similarly, 2004Al and Zn were formed at 500 °C ($T_{process}$ (2004Al) = 470–520 °C) and 410 °C (T_m (Zn) = 375–430 °C) under 30 MPa pressure, respectively. 2004Al (SUPRAL) alloys are superplastically formable alloys, where similar processing conditions as in MG would be sufficient to fill this material into the mold around its processing temperature [58]. Commercial zinc alloys (ZA 12) has 87–88 % purity with additional aluminum and traces of other elements. Therefore, melting range of the zirconium alloy increases to 55 °C [63]. Because KOH reacts with both metal types, samples were taken out of the molds through careful grinding and breaking the excessive silicon mold.

The behavior of these materials under quasistatic loading is shown in Fig. 4.9. The initiation strength of $Zr_{35}Ti_{30}Cu_{7.5}Be_{27.5}$ cellular structure exceeds those of other considered materials; it is almost three orders of magnitude higher than for PE, around thirty times higher than for PEEK, and three times higher than that for 2004Al cellular structure. When normalized by density, the initiation strength difference between Zr-MG and 2004Al is still two times (Table 4.1; [42]).

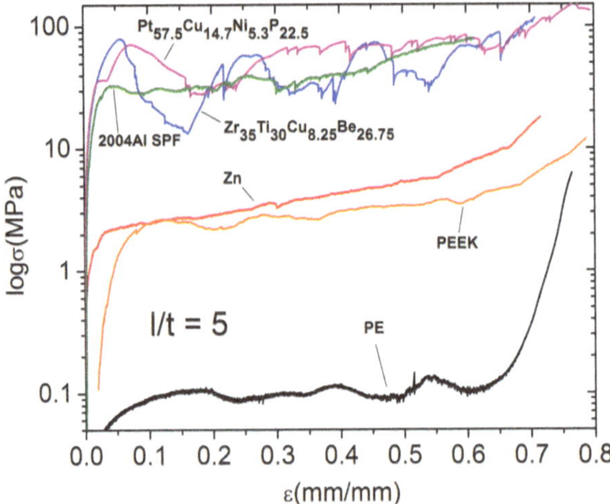

Fig. 4.9 Stress–strain curves of cellular structures with $\rho^*/\rho_s \approx 24.0\,\%$ ($l/t=5$), covering four orders of magnitude fabricated from $Zr_{35}Ti_{30}Cu_{7.5}Be_{27.5}$, $Pt_{57.5}Cu_{14.7}Ni_{5.3}P_{22.5}$, 2004Al, zinc, polyether ether ketone (PEEK), and polyethylene (PE). Note that the control cellular structures are made of much lighter materials (except Zn), which partially explains their lower mechanical properties. (Reprinted from reference [42] with a permission from John Wiley and Sons)

4.2.4.4 Energy Absorption Capacity

According to the experiments conducted in this study, MG-cellular structures are found to be high energy absorbing materials due to the high σ_y together with high plateau strength when $\rho^*/\rho_s < 40.0\,\%$. Figure 4.10 shows the energy absorption per unit volume (W) until densification (ε_D) for different ρ^*/ρ_s ratios of $Zr_{35}Ti_{30}Cu_{7.5}Be_{27.5}$ cellular structures calculated through:

$$W = \int_0^{\varepsilon_D} \sigma d\varepsilon \tag{4.3}$$

where σ and ε represent the instantaneous stress and strain values, respectively.

The distinct deformation modes affect the energy absorption capacities. Even though the initiation strength is very high in structures with densities exceeding 40.0 %, the energy absorption values are lower than for the low-density structures. This is due to the catastrophic failure after reaching σ_I. Below a density of 40.0 %, cellular structure deforms without losing its overall integrity. As a consequence, they can absorb larger energies prior to overall collapse. W reaches a maximum value of ~31.0 MPa at $\rho^*/\rho_s \approx 24.0\,\%$, which is more than twice of its bulk value.

It is found that porosity can be used as an effective design tool to enhance energy absorption mechanism for metallic glasses. For example, W of a cellular structure normalized by $\rho^*/\rho_s \approx 2.5\,\%$ exceeds that of the monolithic MG, however, at only 10 % of its weight (Fig. 4.10 inset). Compared to other MG cellular structures in the literature, the energy absorption measured here for the MG cellular structure represents the upper bound (Fig. 4.11; [42]).

Table 4.1 Mechanical properties of cellular structures from different materials. (Reprinted from reference [42] with a permission from John Wiley and Sons)

$\rho^*/\rho_s = 24.0\,\%$	$Zr_{35}Ti_{30}Cu_{7.5}Be_{27.5}$	$Pt_{57.5}Cu_{14.7}Ni_{5.3}P_{22.5}$	2004Al	Zn	PEEK	PE
σ_I [MPa]	80.9	72.5	33.7	2.2	2.7	0.1
σ_p [MPa]	45.7	60.8	35.2	4.4	3.2	0.1
ε_D [%]	68.4	65.8	41.6	66.5	68.6	64.6
U_T [MPa]	31.0	38.6	15.0	2.8	2.0	6.0E-02
$\Delta\sigma_a$ [MPa]	19.4	15.7	4.3	1.9	0.7	1.0E-02
ε_{el} [%]	5.4	7.5	4.2	0.2	2.7	27.4
E^* [MPa]	1730.0	1840.0	1220.0	103.0	45.0	0.2
ρ [g/cm^3]	3.9	15.0	2.8	7.1	1.3	0.9

Errors are in the order of the last digit shown. $\varepsilon_{el} = \frac{\Delta L}{L_0}$ (elastic strain), $W = \int_0^{\varepsilon_D} \sigma d\varepsilon$ (energy absorption per unit volume), $\sigma_p = \frac{1}{n}\sum_{i=1}^{n} \sigma_i$ (average value of the stress plateau), σ_I = Initiation stress, ε_I = Initiation strain, $\Delta\sigma_a = \sqrt{\frac{1}{N}\sum_{i=1}^{n}(\sigma_i - \sigma_p{}^2)}$ (amplitude of stress undulations (standard deviation of σ_p)), ε_D = Densification strain, $\Delta\varepsilon_P = \varepsilon_D - \varepsilon_I$ (plateau strain), E^* = Modulus of the cellular structure (slope of the linear elastic region)

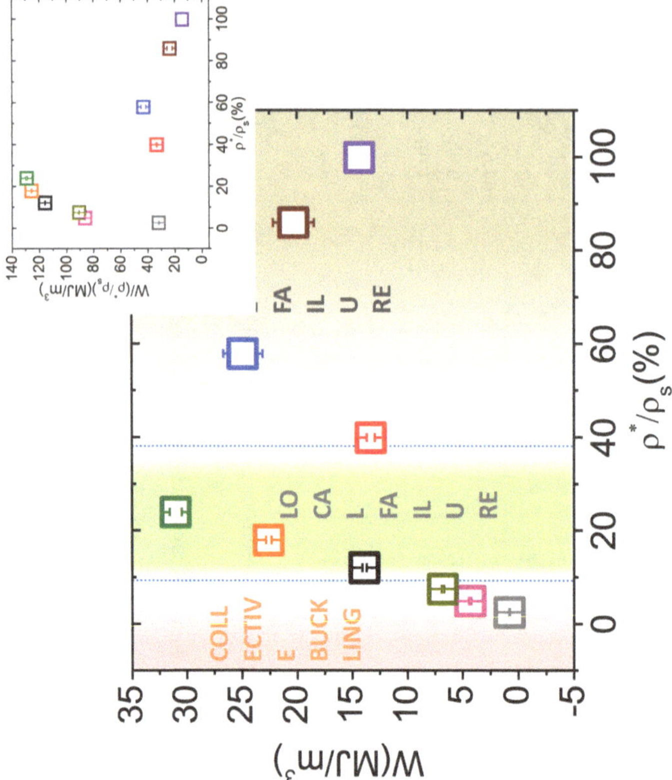

Fig. 4.10 Energy absorption per unit volume as function of density. W increases with increasing density up to ~25.0%, where the best compromise of strength and plasticity is present for $18.0\%<\rho^*/\rho_s<24.0\%$. W decreases after this point due to the change in the deformation mode from localized to global failure. The errors indicate the average value of three tests for W. The inset shows the energy absorption values normalized by the relative density of the cellular structure, where the normalized energy absorption ($W/(\rho^*/\rho_s)$) for $\rho^*/\rho_s\approx 24.0\%$ is around an order of magnitude higher than that of the bulk MG sample. (Reprinted from reference [15] with a permission from Elsevier)

In comparison with the other materials considered in Fig. 4.9, the MG cellular structure ($\rho^*/\rho_s\approx 24.0\%$) absorbs approximately four orders of magnitude more energy than PE, one order of magnitude than PEEK and Zn, and twice that of 2004Al. When energy absorption is normalized by material's density, MG and 2004Al cellular structures are comparable (Table 4.1).

Due to the ability of some of the MG honeycombs to elastically buckle, variations in resilience (elastic energy absorption per unit volume) as a function of ρ^*/ρ_s can be expected. The resilience (U_R) is defined as:

$$U_R = \int_0^{\varepsilon_{el}} \sigma d\varepsilon \tag{4.4}$$

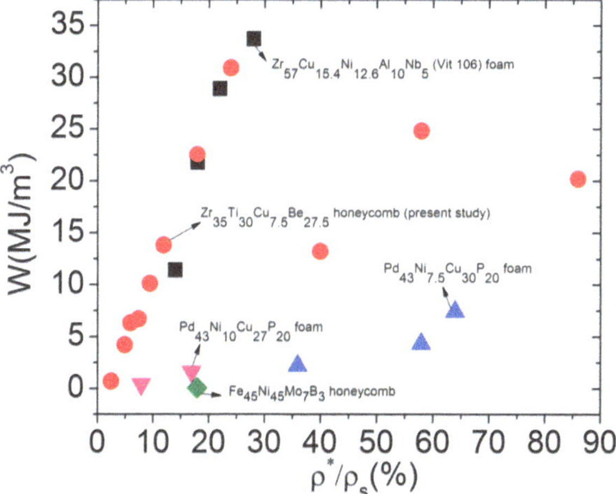

Fig. 4.11 A comparison of energy absorption of MG cellular structures including $Zr_{35}Ti_{30}Cu_{7.5}Be_{27.5}$ cellular structure (present study), $Zr_{57}Cu_{15.4}Ni_{12.6}Al_{10}Nb_5$ open-cell foam, $Pd_{43}Cu_{27}Ni_{10}P_{20}$ foam, and $Fe_{45}Ni_{45}Mo_7B_3$ cellular structure. (Reprinted from reference [42] with a permission from John Wiley and Sons)

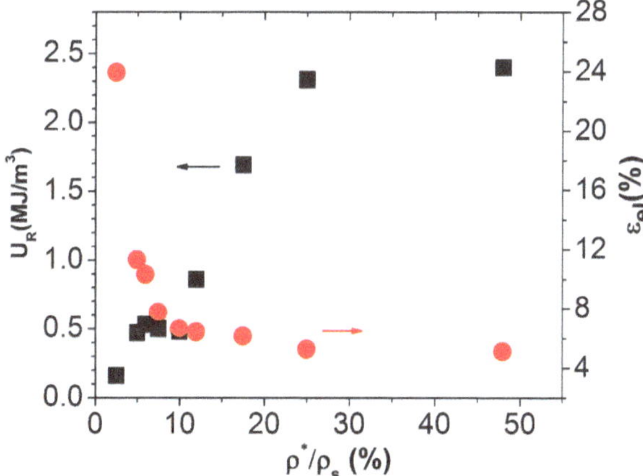

Fig. 4.12 Resilience and elasticity (linear and nonlinear) vs. *l/t*. *Black squares* represent the resilience values, which are calculated according to Eq. 3 using the elastic recovery data (ε_{el}) indicated by the *red square*

Figure 4.12 shows the elastic energy absorbed for different ρ^*/ρ_s. U_R increases in an approximately linear fashion up to a certain density after which the resilience almost stabilizes. On the other hand, total elasticity of the cellular structures follows an exponential decay as ρ^*/ρ_s increases, which is reflected on the elastic energy absorption.

4.2.4.5 Microstructural Optimization

Our artificial microstructure approach allows to identify a hierarchy of microstructural features. The knowledge of the relative and absolute importance of individual features can be used to optimize microstructures. As a design tool to enhance MG cellular structures' performance, we considered cellular structure cell corner-fillets to reduce stress concentrations (see Fig. 4.1d). To study the effectiveness of filleting, we used cell joints of different radii in $\rho^*/\rho_s \approx 5.0\,\%$ MG cellular structures, and considered fillets with radius ranging 1–50 μm (Fig. 4.13). CAD design and two-step replication process of MGs provides an unparalleled control in corner-filleting which cannot be achieved by any other fabrication methods. For a structure with $t = 20$ μm, our results reveal a critical filleting radius (r) of 6 μm above which the strength of the cellular structure increases. Enlarging this radius beyond 6 μm did not enhance the effect on initiation strength further. Our finding of such a critical ratio provides an effective design tool for increasing the strength of MG cellular structures; strength doubled when corner-fillets are introduced at the expense of only 0.2 % density increase. This finding was in excellent agreement with the other cellular structures with different relative densities, therefore, a critical fillet ratio of $r/t = 0.3$ was selected as the design criteria.

For a cellular structure with $\rho^*/\rho_s < 12.0\,\%$, elastic buckling occurs. Because stresses are maximal at the cellular structure cell joints, yielding initiates at these regions. Microstructural analysis is carried out for $Zr_{35}Ti_{30}Cu_{7.5}Be_{27.5}$ cellular structures to reveal the fraction of the MG cellular structure that plastically deforms, which indicates the effectiveness of the structure for energy absorbing applications (Fig. 4.14). Our findings reveal that deformation through shear banding occurs up to the densification limit (ε_D) without an indication of fracture. The shear bands, which carry plastic deformation, are most densely present at the joints of the cel-

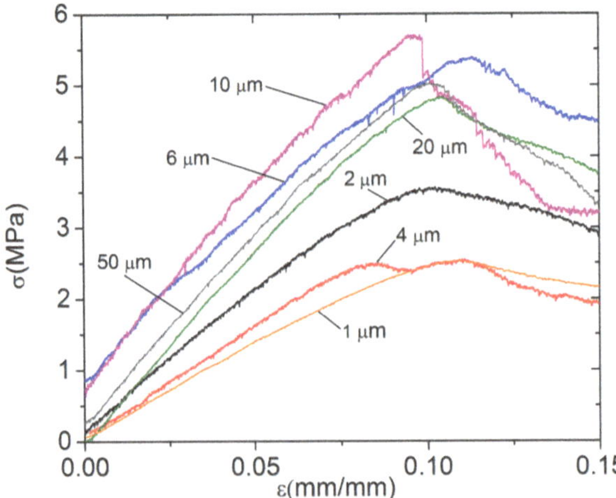

Fig. 4.13 Effect of filleting corners of the cellular structure cell on the initiation strength of $Zr_{35}Ti_{30}Cu_{7.5}Be_{27.5}$ cellular structure ($\rho^*/\rho_s \approx 5.0\,\%$). Significant increase in σ_p is achieved for corner-fillet radius ≥6 μm. (Reprinted from reference [42] with a permission from John Wiley and Sons)

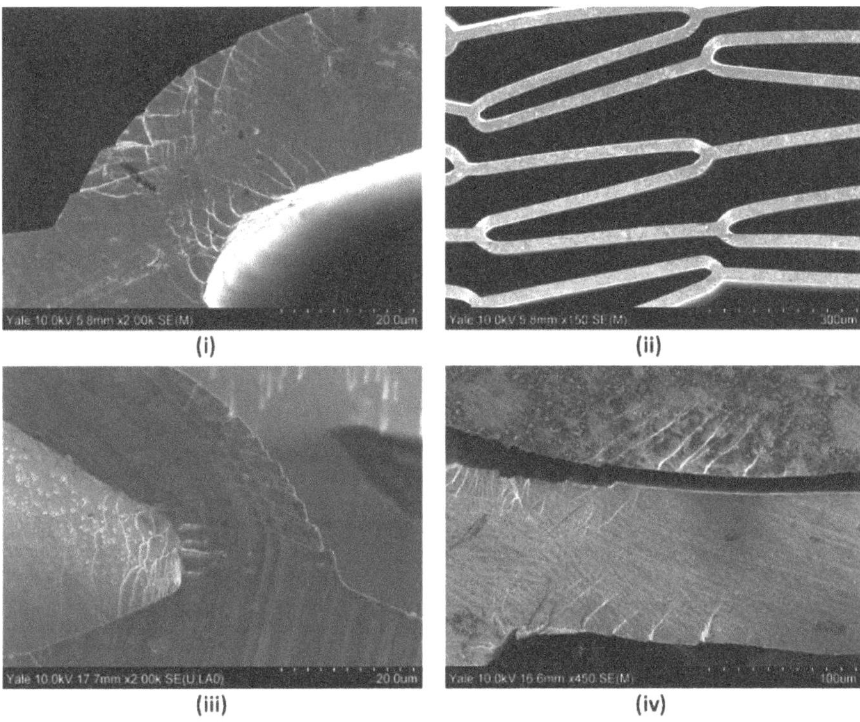

Fig. 4.14 Corresponding SEM images of $Zr_{35}Ti_{30}Cu_{7.5}Be_{27.5}$ MG cellular structures after deforming until ε_D. (*i, ii*) Shear bands are localized at joints for corner fillets with $r<1$ μm ($\rho^*/\rho_s \approx 5.0\,\%$), and no shear band is observed at the ligaments. (*iii, iv*) Shear bands are more homogenously distributed throughout the cellular structure when using corner-fillets with $r>6$ μm with a shear band spacing of ~30 μm ((*iii*) $\rho^*/\rho_s \approx 5.0\,\%$ and (*iv*) $\rho^*/\rho_s \approx 12.0\,\%$, respectively), which spread out up to (*iii*) ~30 μm and (*iv*) ~130 μm. (Reprinted from reference [42] with a permission from John Wiley and Sons)

lular structures. However, in cellular structures with corner-fillets, shear bands distributed more evenly throughout the cellular structure cell, hence utilizing a larger fraction of the cellular structure cell, which results in higher energy absorption (Fig. 4.14(iii) and 4.14(iv); [42]).

To identify geometries where corner-fillets are most effective for $Zr_{35}Ti_{30}Cu_{7.5}Be_{27.5}$ cellular structures, we compared the effects of filleting for various ρ^*/ρ_s ratios. As a measure, two kinds of corner-fillet geometries were considered, $r/t=0.3$ (6 μm for $\rho^*/\rho_s \approx 5.0\,\%$) and $r/t<1$ μm (smallest radius possible with the resolution of photolithography; Fig. 4.15). The effect of filleting on σ_p is most pronounced for $\rho^*/\rho_s \approx 7.5\,\%$ [42].

4.2.4.6 Embrittlement of Metallic Glasses

The properties of MG forming alloys can be dramatically altered by annealing, which results in structural relaxation or crystallization [42]. During annealing, MGs relax into the supercooled liquid, and extended sub-T_g annealing changes the

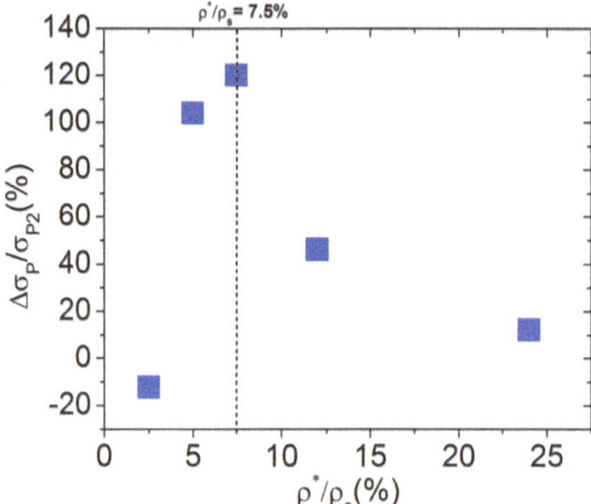

Fig. 4.15 Effects of fillets on the plateau stress of Zr-MG cellular structures as a function of l/t. Data points represent the comparison of the cellular structures with $r/t = 0.3$ and $r < 1$ µm. The effect of corner-fillets is most pronounced for $\rho^*/\rho_s \approx 7.5\,\%$. $\Delta\sigma_p$ is defined by the difference between the propagation strength of the cellular structures with $r/t = 0.3$ (σ_{p2}) and $r < 1$ µm (σ_{p1}). (Reprinted from reference [42] with a permission from John Wiley and Sons)

structural configuration of the atoms, which generally result in severe embrittlement of the material. The embrittled MG sample, where the degradation is highly composition dependent, shows very low bending ductility and fracture toughness compared to its amorphous state [39, 64–68].

To explore the effect of MG properties on the performance of the cellular structures, we annealed the Zr-MG cellular structures at temperatures below and above the glass transition temperature for comparable times with structural relaxation and crystallization. Sub-T_g embrittlement is achieved by annealing the as-formed cellular structure at 275 °C for 18 h, whereas the crystallized cellular structure is formed by heating to 400 °C (above-T_g) for 30 min [39, 68, 69]. Their performances are compared to the as-formed cellular structure of the same ρ^*/ρ_s ratio (Fig. 4.16). The embrittled and crystallized cellular structures show pronounced shear localization and a yield plane close to 45° to the applied load where cell walls fracture (Fig. 4.16b). The as-formed cellular structure, on the other hand, deforms through row collapsing parallel to the applied stress achieved by plastic deformation of the ligaments without fracture. Initiation, propagation strength, and elastic strain limit of the embrittled and crystallized cellular structures are lower than those of the amorphous cellular structure. These findings, together with the comparison of MG cellular structure with cellular structures made of different materials, demonstrate the importance of the material's property for the cellular structure's performance (Table 4.2; [39]).

Table 4.2 shows the theoretical strength (σ_{y_theo}) calculated by equating the maximum moment in the beam to the fully plastic moment of the cell wall in bending conditions [48]:

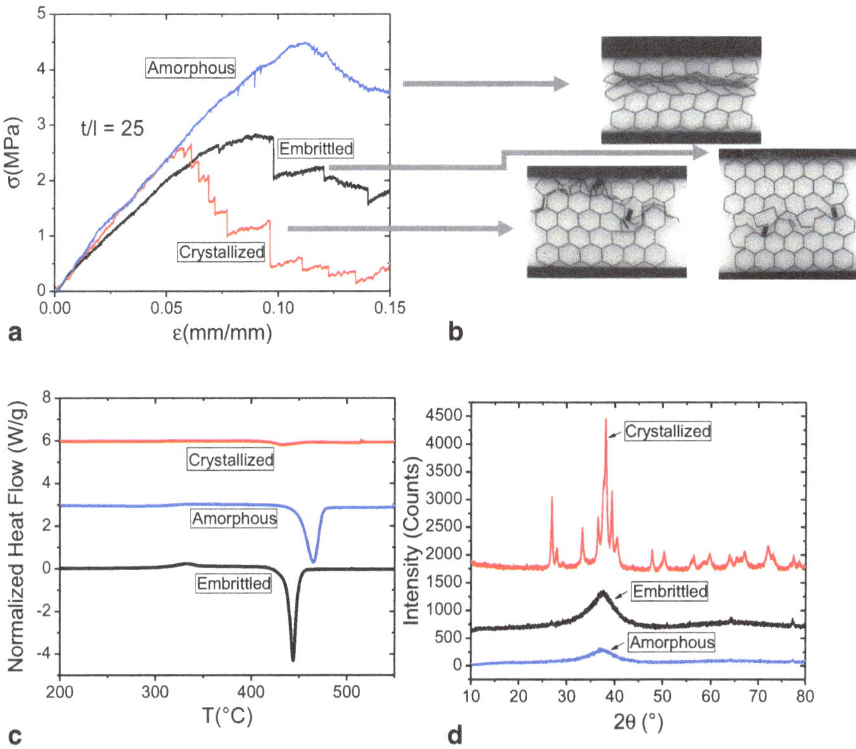

Fig. 4.16 Influence of sub-T_g (embrittlement) and above-T_g (crystallization) annealing on mechanical properties of Zr-MG cellular structures with $\rho^*/\rho_s \approx 5.0\,\%$ (constant). **a** Mechanical properties and **b** deformation behavior of as-formed, embrittled, and crystallized cellular structures are compared. **c** Normalized heat flow and **d** XRD spectrum of the differently processed cellular structures. (Reprinted from reference [42] with a permission from John Wiley and Sons)

$$M_P = \frac{1}{4}\sigma_{ys} b t^2 \text{ (fully plastic moment)} \tag{4.5}$$

$$M_{max} = \frac{3}{4}\sigma_1 b l^2 \text{ (maximum moment in the beam when } \theta = 30^\circ \text{ and } h = l) \tag{4.6}$$

It can be extracted from this table that the real measurements of the initiation strength (σ_I) is always higher than the estimated theoretical values, which can be explained by the microstructural optimization of the joint sections of the MG cellular ligaments. σ_I/σ_{I_theo} decreases from 5 to ~2 as the relative density increases from 2.5 to 24.0 %, which reveals that the stress optimization is most effective when used for cellular structures with lower relative density. The theoretical stiffness (E^*_{theo}) is calculated from the standard beam deflection theory and the moment tending to bend the cell wall [48]:

Table 4.2 Mechanical properties of MG cellular structures for different l/t ratios. Errors are on the order of the last digit shown. $E^{*}_{theo} = 2.3E\left(\frac{t}{l}\right)^3$ (calculated theoretical modulus of the cellular structure), $\sigma_{f\_theo} = \frac{2}{3}\sigma_y\left(\frac{t}{l}\right)^2$ (calculated theoretical initiation strength of the cellular structure). 20_cryst & 20_emb: crystallized and embrittled sample (l/t=20). (Reprinted from reference [42] with a permission from John Wiley and Sons)

l/t	50	25	20	16.7	12.5	10	6.7	5	2.5	20_cryst	20_emb	Bulk [25]
ρ^*/ρ_s [%]	2.4	4.7	5.8	6.9	9.2	11.6	17.3	23.3	46.8	4.7	4.7	100
ε_{el} [%]	24	11.4	10.4	7.8	6.7	6.5	6.2	5.4	5.2	1.00E-002	0.1	2
W [MPa]	0.8	4.3	6.4	6.8	10.2	13.9	22.6	31	64.8	1.2	1.8	14.3
σ_p [MPa]	0.9	5.2	8.4	8.6	13	18.8	31.2	45.7	108.3	1.48	2.36	
σ_I [MPa]	1	5.4	7	10.1	12.2	22.3	49.2	80.9	301.2	2.7	2.8	1430
$\sigma_{I\ theo}$ [MPa]	0.2	1.3	2.5	3.5	6.3	9.8	22.1	39.2		1.3	1.3	
$\Delta\sigma_a$ [MPa]	0.2	1	1.3	1.4	3.1	5.1	10.5	19.4	46	0.7	0.3	
ε_D [%]	85.4	84.2	78.3	81.4	81.7	75.8	72.2	68.4	64.8	82.3	77.1	
E^* [MPa]	8.9	46	61	130	215	365	920	1730	5600	46	38	
E^*_{theo} [MPa]	1.6	12.8	25	43.2	102.3	199.9	674.6	1599		12.8	12.8	86,900

$$M = \frac{Pl}{4} \tag{4.7}$$

$$P = \frac{3}{2}\sigma_2 lb \tag{4.8}$$

$$\delta = \frac{Pl^3}{24EI} \tag{4.9}$$

$$\varepsilon_2 = \frac{\delta}{\sqrt{3}l} = \frac{\sigma_2 bl^3}{16\sqrt{3}EI} \tag{4.10}$$

$$E_2^* = \frac{\sigma_2}{\varepsilon_2} \tag{4.11}$$

E^*_{theo} is found to be smaller than the measured stiffness (E^*). The difference between these two values gets smaller as the relative density increases, which can also be correlated with the structural optimization of the cell joints.

The deformation behavior of the cellular structures, as discussed above, is linked to the standard beam deflection of the ligaments, and the plastic deformation at the joint sections. Therefore, one can estimate the deformation behavior of a variety of materials because the geometrical effect of deformation is correlated with the internal properties as discussed in Eq. 4.2. An equation can be derived using the formulas given (4.9) and (4.10), by which the elastic modulus of any hexagonal cellular structure can be estimated:

$$\frac{E_1^*}{E_s} = \frac{E_2^*}{E_s} = \frac{4\sqrt{3}}{3}\left(\frac{t}{l}\right)^3 \tag{4.12}$$

Similarly, derivation of these formulas in Eqs. 4.5–4.12 can also be manipulated for calculating in-plane stiffness and yield strength of other periodic metal honeycombs with respect to the geometry of the cells [48].

4.2.4.7 Comparison with Numerical Simulations

Based on the equation given above, finite element modeling analysis has been conducted for the in-plane compressive deformation of an aluminum honeycomb sample with $\rho^*/\rho_s \approx 3.0\,\%$. The results indicate that numerical simulations of the selected cellular structure reproduced all the major deformation characteristics of the real deformation of this honeycomb carried out simultaneously (Fig. 4.17; [70]). σ_I was found to be higher than the experimentally measured data which can be linked to the small geometric imperfections within the structure. On the other hand, σ_p is showing excellent agreement with the experimental data and the location of the stress

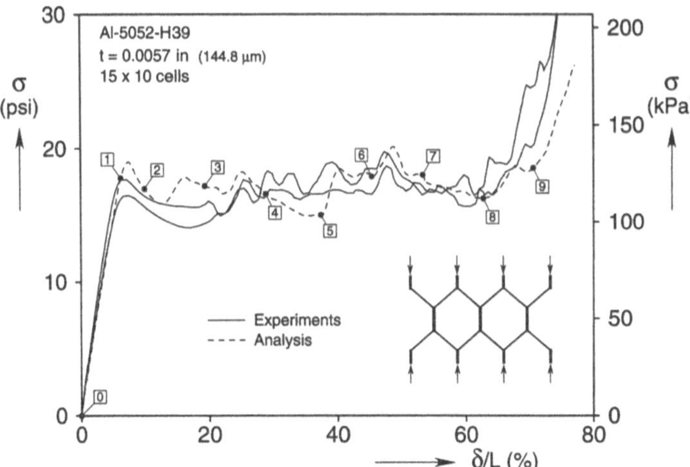

Fig. 4.17 Two experimental results compared to the numerical simulation data for $\rho^*/\rho_s \approx 3.0\,\%$ cellular structure. (Reprinted from reference [71] with a permission from Elsevier)

plateaus. There is a negligible difference for the estimation of the stress fluctuations ($\Delta\sigma_a$), which confirms the accuracy of the simulation techniques. In general, FE simulations have found to be reliable tools to estimate the properties of cellular structures, particularly for the ones with low ρ^*/ρ_s ratio where the cells deform row-by-row and remain intact throughout the deformation.

4.2.4.8 Mechanical Characterization Under Uniaxial Tension

Compressive fracture is insensitive to defects like flaws, cracks, or a few exceptionally large cells. But this is not true for tensile fracture. Since all materials include imperfections which induce stress concentrations, crack is initiated after exceeding the yield strength of the cellular structure. This leads to sudden fracture of the ligaments and catastrophic failure.

Stress-strain curves of $\rho^*/\rho_s \approx 9.5\,\%$ $Zr_{35}Ti_{30}Cu_{7.5}Be_{27.5}$ cellular structures under in-plane tensile and compressive deformations are compared in Fig. 4.18. Unlike the low yield strength followed by smooth stress serrations observed in compressive deformation of cellular structures, the uniaxial tension creates a stretching effect on the ligaments and results in a linear-elastic curve until the sample fractures from one cell joint, which causes a sharp decrease in stress value. Stress fluctuations are observed after this point due to individual ligament fractures as the crack propagates across the cellular structure. Despite the σ_I is ~50 % higher for tensile sample due to this stretching effect, early fracture of the ligaments causes W to be only ~40 % of the compression sample of the same relative density (9.5 %).

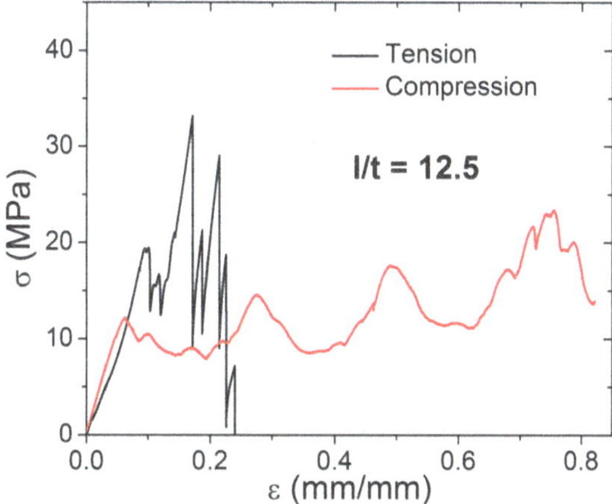

Fig. 4.18 Stress-strain curve for tensile and compressive deformation of Zr-MG cellular structure with the same ρ^*/ρ_s ratio of ~9.5 % (l/t = 12.5). Tensile forces acting on the cellular structure causes early crack initiation, which triggers the crack growth across the cross-section and thereby, the failure happens at an earlier stage compared to the compression mode

4.2.4.9 Mechanical Characterization at Different Orientations

Figure 4.19a shows the comparison of the hexagonal cellular structures characterized in two different orientations. The experimental results show that for regular cellular structure cells with $\theta = 30°$, σ_I in the X_1 and X_2 directions is very similar as suggested in [71]. However, σ_p and σ_a values are higher for the deformation in X_2 direction, resulting in ~20 % higher W. This can be due to the contribution of the high amount of nonlinear elastic buckling of the vertical ligaments in MG cellular structures in X_2 direction prior to plastic deformation of each row. Since there are no vertical columns distributing the initial loading to the same row in X_1 direction, unlike the row-by-row collapse observed for the X_2 direction, the deformation for the $\rho^*/\rho_s \approx 9.5$ % sample occurs at an angle where the resolved shear stress is maximum (Fig. 4.19b). For this reason, to obtain better mechanical properties, in-plane compression of all the cellular structures in this chapter was conducted at X_2 orientation.

4.2.5 General Findings and Conclusions

By artificial microstructure approach, we introduced an effective method to manipulate MG honeycombs. Characteristics of the cellular structure such as the ligament length, thickness, and radius of curvature at the joints of the cells are varied all individually to determine how changes in these characteristics affect properties under uniaxial in-plane compression test. We found that the deformation behavior of MG cellular structures can be controlled through microstructural design, from brittle to ductile, by changing the relative density of the cellular structure through changing length to thickness ratio of the ligaments. The deformation of cellular Zr-

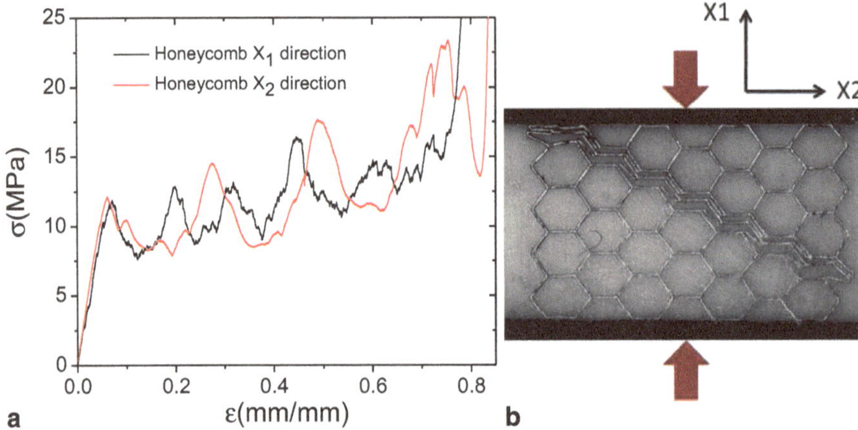

Fig. 4.19 **a** Stress-strain curves of cellular structures (*hexagonal*) at different orientations. **b** Deformation is highly localized on the direction having the highest stress concentration when the sample is compressed along X_1 direction

MGs with a wide range of relative densities from 2.5 to 86.0% was investigated, and three distinctive deformation regions are revealed. With increasing density, the deformation changes from collective buckling with high elasticity and plasticity at the expense of low strength ($\rho^*/\rho_s < 12.0\%$) to local failure causing shear band-like fracture with high strength and plasticity ($12.0\% < \rho^*/\rho_s < 40.0\%$). Global failure with high strength but negligible plasticity is the dominating deformation mechanism when $\rho^*/\rho_s > 40.0\%$. The optimal density for energy absorption of ~25.0% represents the best compromise of strength and plasticity, and thereby, energy absorption capacity in the local failure region.

Furthermore, we identified cell corner-fillets as effective design features; a 0.2% increase in density doubles strength and energy absorption. Results also have revealed that MG cellular structures absorb energy more than twice as much in the bulk monolithic form and exceeding cellular structures of most other materials. This is because in the cellular structure a size effect is utilized, which promotes plastic deformation [15].

4.3 Toughening Mechanisms in MGs

As a second example in the material class of MGs, we used our artificial microstructure approach to understand toughening mechanism in MG heterostructures. One emerging strategy to couple the attractive properties of MGs with plasticity is to introduce a second phase, which reflects and/or absorbs shear bands. It has been suggested that this strategy should be particularly successful when the spacing of the second phase coincides with the plastic zone size of the MG matrix phase [3, 10, 20, 35, 37, 72–76]. However, verifying this hypothesis as the origin of the toughening mechanism has been challenging since the fabrication methods of the MG-composites and foams are complex, and do not allow independent and systematic variation of microstructural features such as phase spacing, size, shape, and volume fraction. We will use our artificial microstructure strategy and vary the feature aspects independently, which will allow us to quantify their individual contributions.

4.3.1 Uniaxial Tensile Test

Mechanical characterization of the MG heterostructures was conducted using Instron 5543 Tensile Tester with 1 kN maximum load capacity under quasistatic conditions (strain rate of 0.001 s^{-1}). A pair of flat steel molds was used to pin the MG heterostructures from the top and bottom parts through four pinholes on each side. The sample was placed vertically on the loading direction, and the tensile grips of the Instron held the steel molds tight by pneumatic (air) pressure. Slack correction was used to eliminate the errors caused by sample positioning and instant loading at the beginning of the test. An extensometer having a gauge length of 8 mm was utilized to measure the exact displacement of the sample during deformation. Stress–strain curve data were retrieved by using the software tool of Instron, where the plots were subsequently generated by Origin 8.5 graphing program [43].

4.3.1.1 Effect of Pore Size

Tensile stress–strain curves for various MG heterostructures (MG and pores), and as a comparison for a monolithic MG sample, are shown in Fig. 4.20. The monolithic MG fabricated with our method exhibits an elastic strain limit of 2 % and yield strength of 1750 MPa, which is almost identical to the elastic strain limit of 2 % and yield strength of 1765 MPa, typically measured in the as-cast state of bulk monolithic rod-shaped samples in tension [43]. These almost identical values confirm that our fabrication method itself does not affect the mechanical properties. MG heterostructures with specific microstructural architecture exhibit a remarkably

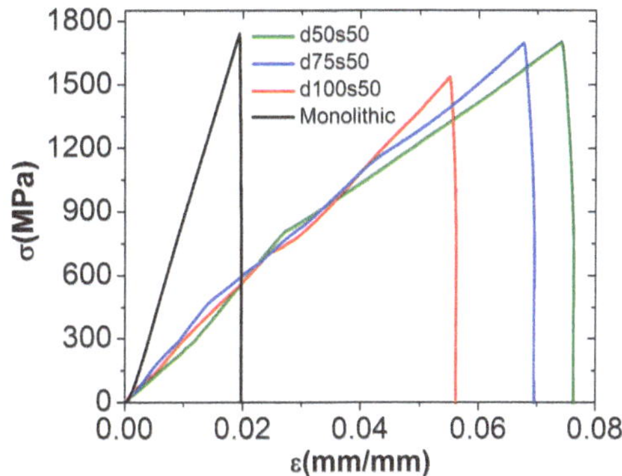

Fig. 4.20 Quasistatic tensile characterization of MG heterostructures with various pore sizes compared to monolithic MG. Nominal fracture strength of the samples is calculated from the effective area, where the effective width is the sample size minus the size of the pores [43]

different mechanical response. For a circular microstructural architecture with pore diameter (d) and spacing (s) of 50 μm, the fracture strain increases dramatically to $\varepsilon_f \approx 7.5\,\%$. At the same time, Young's modulus of the heterostructure decreases, whereas the nominal fracture strength, σ_f, remains relatively unaltered [77].

4.3.1.2 Effect of Morphology

To deconvulate elastic and plastic deformation, we carried out cyclic loading with gradually increasing loads (Fig. 4.21a). A heterostructure with $d = 100$ μm and $s = 50$ μm was selected, which is much easier to fabricate than a $d = s = 50$ μm heterostructure due to TPF conditions and stress concentration cracking of the Si features while etching. We found that of the 5.5 % strain to failure, ~70 % (3.9 % overall) corresponds to elastic and ~30 % (1.6 % overall) corresponds to plastic deformation. The large elastic deformation originates from the AB pore stacking order. A heterostructure with the same density (same pore size and spacing) but AA-stacking exhibits a significantly lower strain to failure of 2.6 %, where only ~54 % (1.4 % overall) deformation is elastic (Fig. 4.21b; [43]). Table 4.3 compares the amount of elastic and plastic deformation at several unloaded conditions. Total plasticity increases linearly with increasing fracture strain (ε_f), where the AB-stacking morphology creates a geometric effect which allows to reach such high-strain values.

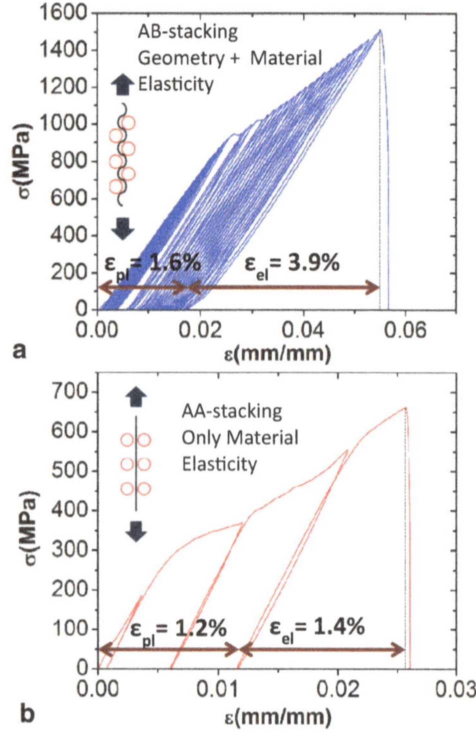

Fig. 4.21 Increased cyclic loading of MG heterostructures with $d = 100$ μm and $s = 50$ μm, and the corresponding engineering stress–strain data. **a** AB-stacking reveals ~3.9 % elastic deformation and ~1.6 % plastic deformation. **b** The amount of deformation in AA-stacking is ~1.4 % elastic and ~1.2 % plastic [43]. Change in the slope after the applied stress reaching to a certain level in both cases is due to multiple shear band formation around the same stress level after this point, where these bands carry the plastic deformation until fracture

Table 4.3 Distribution of plastic to elastic strain under cyclic loading at different strain values for AB-stacking. The percentage of plastic deformation ($\%\varepsilon_p$), which increases linearly until fracture, gives an idea of the plastic strain at any point of deformation. The last row (shaded in *pink*) represents each strain value at the last cycle of AA-stacking

$\%\varepsilon_f$	$\%\varepsilon_p$	$\%\varepsilon_{el}$	$\%\varepsilon_p$
2	0.2	1.8	10
4	0.8	3.2	20
5.5	1.6	3.9	29
2.6	1.2	1.4	46

4.3.1.3 Effect of Pore Spacing

When comparing MG heterostructures of identical pore size, but increasing pore spacing from 25 to 100 μm, both σ_f and ε_f increases. However, when the spacing exceeds 100 μm, both values drop dramatically (Fig. 4.22). The highest performance in terms of σ_f and ε_f is present in heterostructures with a spacing between 100 and 200 μm. This spacing coincides with the plastic zone ahead of a crack tip, $R_p \approx 159$ μm, of $Zr_{35}Ti_{30}Cu_{7.5}Be_{27.5}$. It has been previously stated that [3] plasticity can be generated in the material as long as R_p exceeds the thickness of the sample (in our case it is the spacing (s) between two closest pore) under tension. For a Mode 1 opening crack, this value is calculated according to [3, 78]:

$$R_P = \frac{1}{2\pi}\left(\frac{K_{1C}}{\sigma}\right)^2 \quad (4.13)$$

with $\sigma = 1742$ MPa (fracture strength of the monolithic MG measured). K_{1C} value is taken from another medium-range Zr-MG as ~55 MPa $m^{1/2}$ [64].

Figure 4.23 illustrates the CAD drawings of the gauge sections of MG heterostructures with individual variation of pore diameter and spacing.

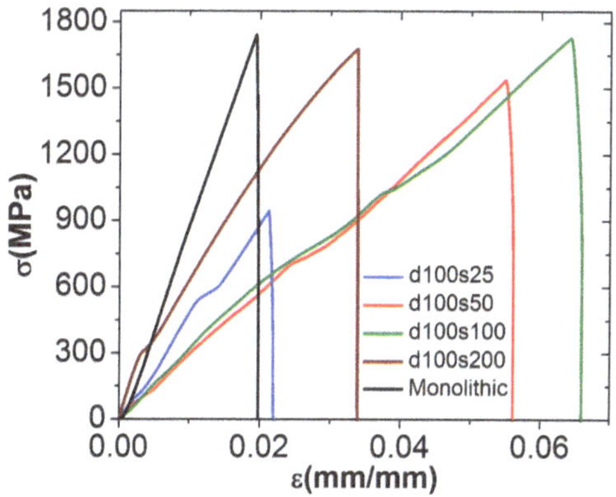

Fig. 4.22 Quasistatic tensile characterization of MG heterostructures with various pore spacing. Tensile ductility and strength increases with increasing pore spacing until this spacing exceeds the plastic zone size [43]

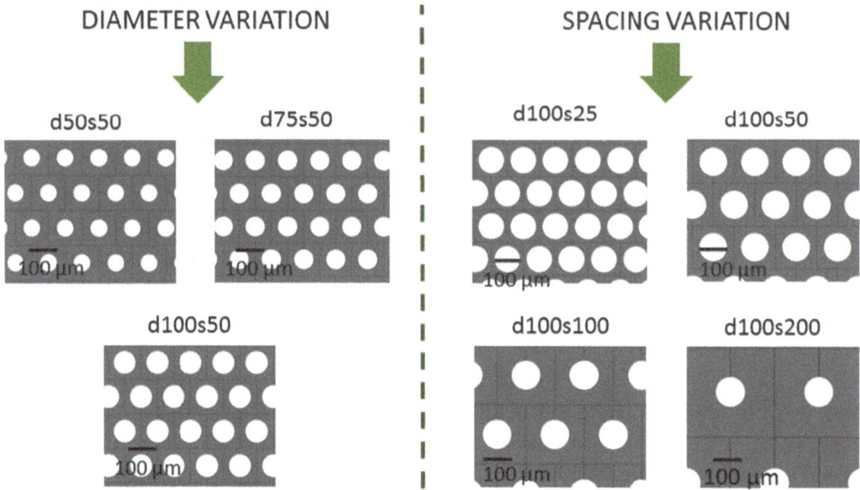

Fig. 4.23 Variation of diameter and spacing as the other parameters kept constant

4.3.1.4 Effect of Pore Shape

Stress concentration around the pores can be manipulated by changing the shape of the pores (Fig. 4.24). Although diameter and spacing of the pores are similar, approximately two times higher strain value and ~20 % higher fracture strength is observed in circular pores due to high-stress concentrations on the sharp tips of microcrack/oval pore features (see Fig. 4.24 inset). This effect is reflected on shear band distribution between the pores, where uniform shear band distribution is observed in circular pores, unlike the heterostructure with microcrack/oval pores which breaks along single shear plane with showing only a few shear bands between neighbor pores perpendicular to the loading direction before fracture.

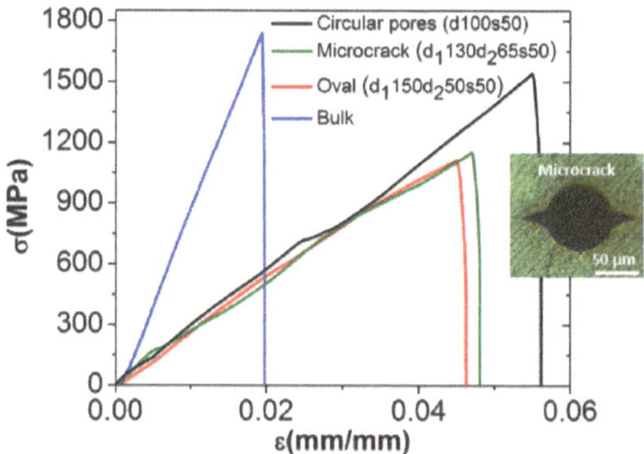

Fig. 4.24 Comparison of horizontal oval pores with 150 µm major diameter along the deformation axis, circular pores having notched tips (maximum diameter of the microcrack: 130 µm-inset), and regular circular pores with 100 µm in diameter

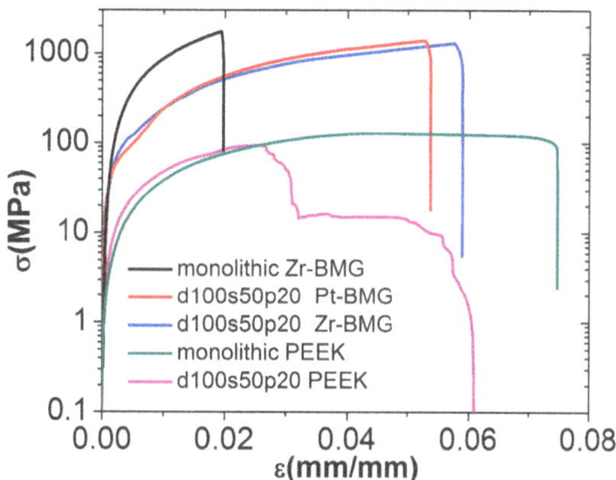

Fig. 4.25 Stress–strain diagram of heterostructure tensile samples ($d = 100$ μm and $s = 50$ μm) fabricated from MGs (Zr and Pt-based), and PEEK. As a reference, monolithic Zr-MG and PEEK samples were compared to their heterostructures

4.3.1.5 Effect of Material Type

One main criterion for the selection of these heterostructures is their suitability for our fabrication process. $Zr_{35}Ti_{30}Cu_{7.5}Be_{27.5}$ sample outperforms $Pt_{57.5}Cu_{14.7}Ni_{5.3}P_{22.5}$ and PEEK samples in terms of fracture strain (Fig. 4.25). Overall toughness of the Zr-MG heterostructure is 10% and ~20 times higher than the Pt-MG and the PEEK heterostructures, respectively. In addition, the specific strength of the Zr-MG heterostructure (σ divided by the density of the related material) is three times bigger than the Pt-MG one. However, it is noteworthy to mention that the reason why Pt-MG has a lower fracture strain might be because the casting diameter of this MG rod is only 2 mm, and hot-rolling at its processing temperature is used to bring the MG into the desired width in multiple sidewise passes, which can induce nanocrystallinity due to exceeding its processing window. Unlike MGs, AB-stacking of pores has an adverse effect on PEEK, resulting in dramatic stress drops after reaching to a certain strain. Therefore, periodic pore alignment is a particular technique for MGs creating multiple shear bands and delay in fracture.

4.3.1.6 Effect of Pore Number

Figure 4.26 shows the effect of change in pore number in a row. Pore numbers of 20 and 30 are selected for comparison. The sample with the number of 30 pores is distributing the applied stress more homogenously across the pores, which increases the ultimate strength. This is because when a shear is initiated between two pores, this region cannot carry the tensile load, and the stress is dissipated between the neighboring pores in the same row. Therefore, increasing the number of pores up to a certain point is predicted to reduce the amount of stress drop during deformation.

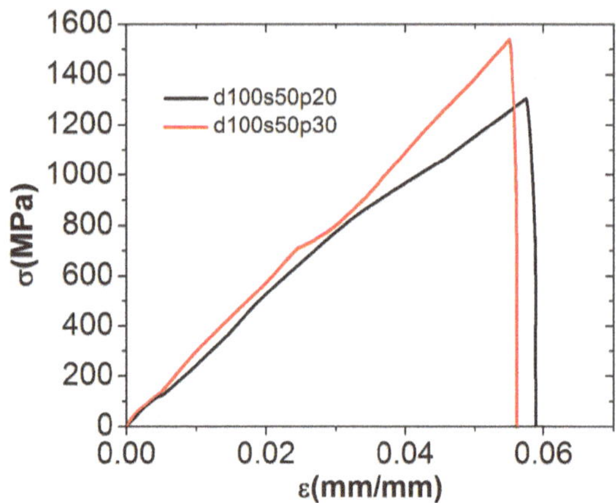

Fig. 4.26 Pore numbers of 20 and 30 are selected, where the sample with the higher number of pores show higher σ_f due to more homogenous stress distribution across the cross-section

4.3.1.7 Effect of Electroplating

To understand the effect of coating on the mechanical properties, the Zr-MG heterostructure sample was electroplated with Ni at a formerly optimized current rate and duration (90 mA, 1 h). Uniaxial tensile test shows that electroplating has an effect on improving the fracture strain by ~10% at the expense of ~10% decrease in σ_f. The coating surface is uniform with a thickness varying between 3–5 μm (Fig. 4.27). This finding shows that Ni coating has a subtle improvement on the overall plasticity. SEM images (Fig. 4.28) confirm that Ni coating provides a shield for some pores by absorbing some of the deformation energy and facilitating the pores to elongate on the deformation axis.

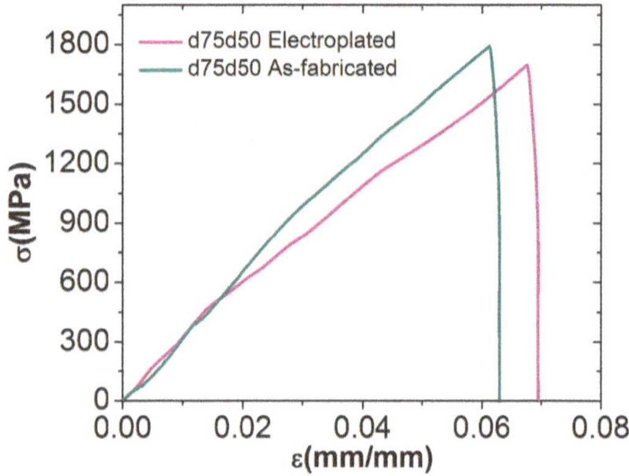

Fig. 4.27 Electroplated sample shows higher fracture strain compared to as-fabricated heterostructure with the same *d*/*s* ratio

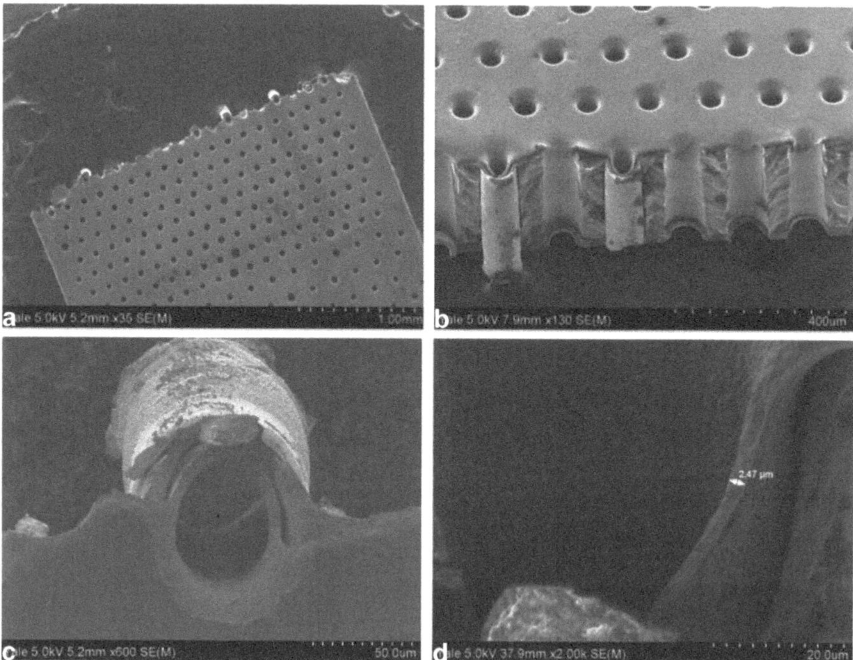

Fig. 4.28 Electrocoated tensile heterostructure with $d=75$ μm and $s=50$ μm. **a**, **b** Some of the pores along the fracture surface perpendicular to the loading direction remain intact as the sample fractures along the horizontal plane due to the contribution of the coated Ni on the elongation mechanism. **c** As a result, some of the Ni coated pores remain intact and become oval-like by tensile deformation. **d** The thickness of the coating shown is ~3 μm, which is detached from MG heterostructure feature after the fracture. SEM images, together with the stress–strain curve in Fig. 4.27, confirm that electrocoating creates an additional reinforcement

4.3.1.8 Microscopic Analysis of the Deformation Mechanism

To understand the microscopic origin of the MG heterostructures' tensile ductility, we carried out SEM imaging (Fig. 4.29). For the $d=s=50$ μm heterostructure which was deformed until failure ($\varepsilon_f \approx 7.5\,\%$), a large number of shear bands form between the pores throughout the entire gauge length (Fig. 4.29a). These shear bands carry the plastic deformation, and their large number results in observed global tensile ductility. Shear bands initiate at pores, where they follow the direction toward the neighboring pore, in a similar pattern like observed in double-notched MG samples (Fig. 4.29b; [79]).

COMSOL Multiphysics Modeling and Simulation Software was utilized to conduct structural element analysis for AB-stacking MG heterostructures. Linear elastic material model was used to approximate the mechanical behavior of the Zr-MGs. The thickness of the specimen was set to 300 μm. The surface was divided into refined triangular meshes, and an applied stress of 1000 MPa is evenly distributed among the pores. von Mises stress, shear stress, and elastic strain around

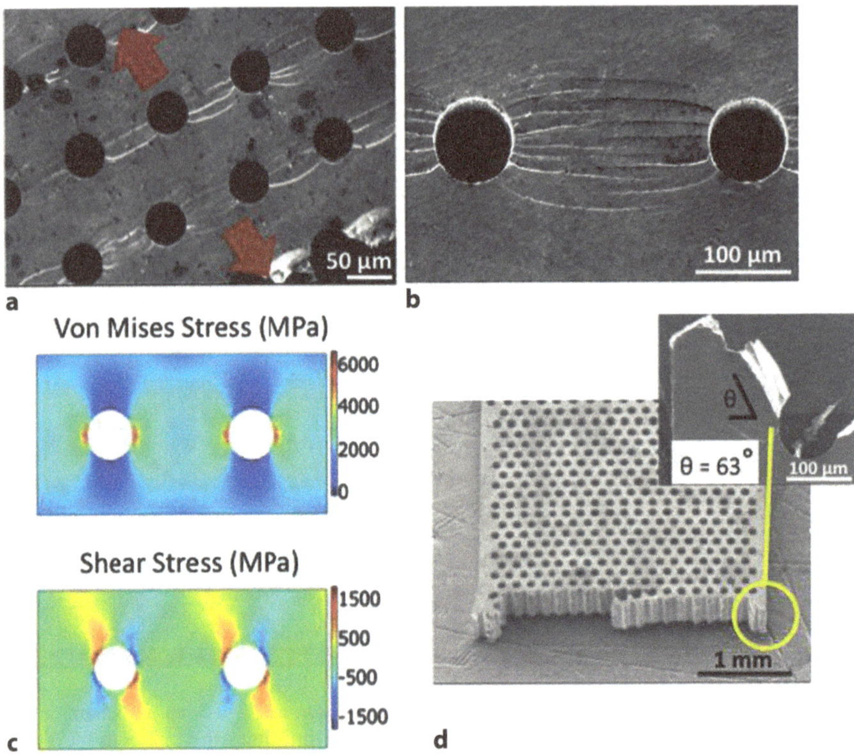

Fig. 4.29 Microscopic analysis of the deformation mechanism in MG heterostructures. **a** Global tensile ductility originates from the formation of multiple shear bands throughout the sample. **b**, **c** Shear concentrations at the pores initiate the formation of shear bands, and redirect their propagation towards the nearest neighbor pore. **d** Fracture of the MG-heterostructure occurs perpendicular to the loading direction. Fracture is redirected at the sample's edges, following the shortest distance between the pores (inset) [43]

the pores were measured and compared with AA-stacking MG heterostructure, and other AB-stacking MG heterostructures with a second phase such as polymer, soft metal, and graphite. Finite element modeling (FEM) simulations show that, as deformation proceeds, the complex stress field (Fig. 4.29c) redirects shear bands towards neighboring pores. Fracture occurs typically along the shortest distance between pores, which is for the considered sample dimensions, perpendicular to the uniaxial loading direction (Fig. 4.29d). At the edges of the sample, fracture chooses a different path by following the shortest distance in the diagonal pore direction (inset in Fig. 4.29d).

It should be mentioned that the magnitude of the stress concentrations depends also on the shear modulus of the second phase, where a second phase with significantly different shear modulus than the matrix causes high-stress concentrations (see Figs. 4.33, 4.34, and 4.35). From FEM results we conclude that the lowest stress concentrations are present for a second phase with a modulus similar to the matrix. However, in this situation, elastic strain values as shown in Fig. 4.21a cannot be achieved.

4.3.1.9 Mechanical Property Optimization Through d/s

The experimental results (Fig. 4.20 and 4.22) reveal the pores' diameter to spacing ratio (*d*/*s*) as the critical feature which controls the mechanical behavior of MG-heterostructures (Fig. 4.30a). Fracture strength decreases with increasing *d*/*s*. Fracture strain, on the other hand, exhibits a maximum at $d/s \approx 1$. The maximum in ε_f originates from an increase in the elasticity and plasticity of the AB-stacking heterostructure with increasing *d*/*s*, which is eventually overcompensated by rapidly increasing stress concentration between pores. As a consequence, the toughness of the MG heterostructure ($U_T = \int_0^{\varepsilon_f} \sigma d\varepsilon$) exhibits a maximum at $d/s=1$ of $U_T \approx 72$ MJ/m^3, which is 4.5 times higher than for the monolithic $Zr_{35}Ti_{30}Cu_{7.5}Be_{27.5}$ MG of $U_T \approx 16$ MJ/m^3 (Fig. 4.30b). It is noteworthy to state that the change in *d* and *s* have negligible effect on fracture stress and strain for the same $d/s=1$ sample. Therefore, we concluded that *d*/*s* ratio is the prominent factor in controlling the stress distribution between the pores, and thereby, the fracture stress and strain.

To reveal the microscopic origin of the dependence of U_T, ε_f, and σ_f on *d*/*s*, we investigated the shear band propagation and distribution across the pores. Specifically, we characterized shear band patterns for different *d*/*s* ratio, and compared those with stress fields, which we determined through FEM. For $d/s=1$, multiple shear bands form throughout the sample (Fig. 4.30c). FEM analysis indicates that resultant von Mises stress in the middle region between the pores perpendicular to the loading direction of approximately 800 MPa is lower than the applied stress of 1000 MPa. Stress concentrations are confined to small, nonpercolating regions adjacent to the pores (Fig. 4.30d). For $d/s=4$, only a very small number of shear bands form prior to fracture (Fig. 4.30f). Thus, the resulting toughness of the heterostructure is low (Fig. 4.30b). FEM results reveal that resultant von Mises stresses for $d/s=4$ are significantly higher (~3200 MPa) for an applied stress of 1000 MPa, and stress concentrations overlap between pores (Fig. 4.30g). As a consequence, MG heterostructures with $d/s=4$ yield at a lower applied load level, and hence, the total amount of plastic deformation until fracture is much smaller (see Fig. 4.22). Thus, only a small number of shear band form before they develop into cracks. The mechanical properties of tensile heterostructures with various *d*/*s* ratios are summarized in Table 4.4. An order of magnitude change in the fracture strain of the heterostructure can be deduced from the table as the sample varies from monolithic state to $d/s=1$ at the expense of a negligible stress drop [43].

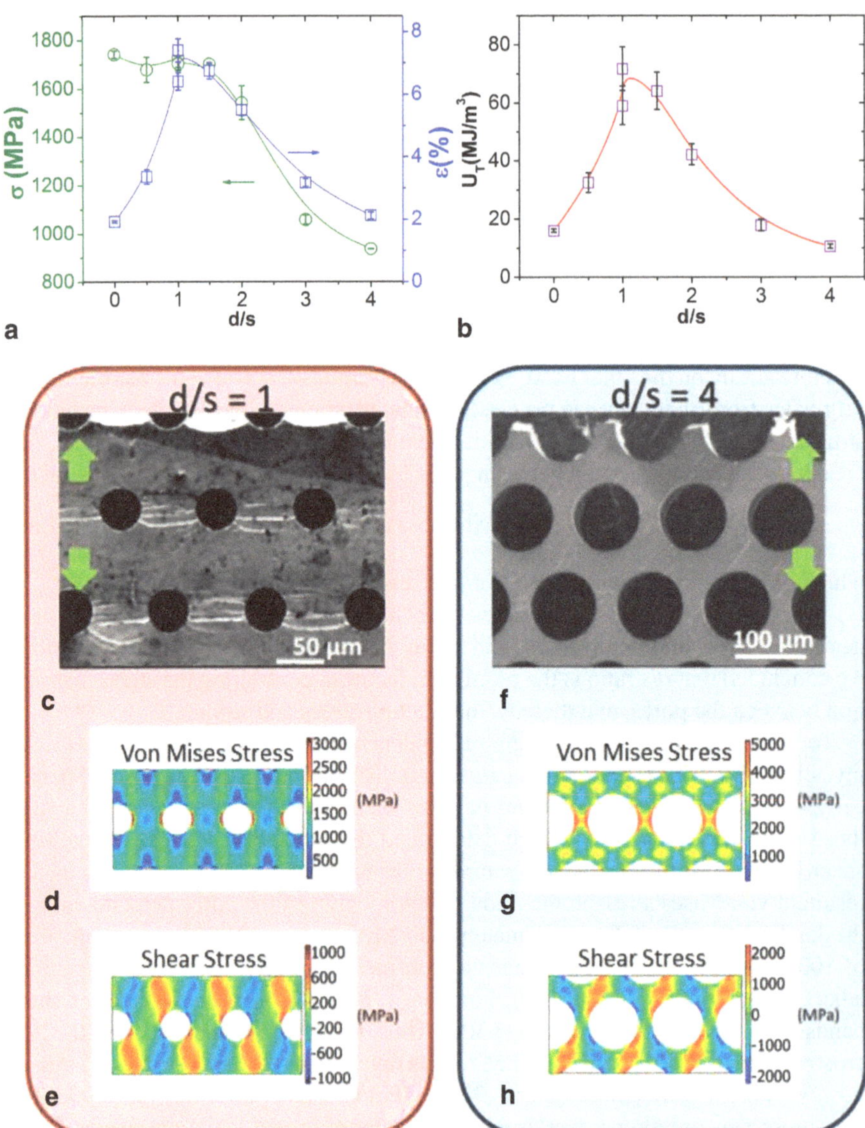

Fig. 4.30 Effect and origin of *d*/*s* on mechanical properties, and comparison between stress fields and shear band patterns in MG heterostructures with different *d*/*s* values. **a** Fracture stress and fracture strain as a function of the ratio of diameter to spacing of the pores (*d*/*s*). **b** Toughness (U_T) vs. *d*/*s* exhibits a pronounced maximum at *d*/*s* ≈ 1. Error bars represent the standard deviation of the mean values from at least three test samples. **c** Corresponding SEM image of MG heterostructures with *d*/*s* = 1 showing multiple shear bands. **d** For an applied average stress of 1000 MPa, the spatially confined stress concentrations between the pores perpendicular to the loading direction are lower compared to heterostructures with *d*/*s* = 4, hence multiple shear bands form without crack formation. **e** Shear bands initiate at the shear stress concentration regions adjacent to the pores. **f–h** For *d*/*s* = 4, stress concentration pathways between the pores already exceed σ_y for the same average applied stress, resulting in early fracture from only a few shear bands [43]

Table 4.4 Mechanical properties of MG heterostructures as a function of *d/s* compared to MG bulk monolithic sample

d/s	σ_{max} (MPa)	ε_{max} (%)	E_{avg} (GPa)	U_T (MJ/m^3)
0 (monolithic)	1742	1.94	94.3	16.0
0.5 (d100s200)	1679	3.37	47.7	32.5
1 (d50s50)	1704	7.41	22.6	71.8
1 (d100s100)	1732	6.42	24.8	59.1
1.5 (d75s50)	1702	6.76	23.8	64.2
2 (d100s50)	1542	5.50	27.2	42.2
3 (d75s25)	1059	3.18	36.5	17.9
4 (d100s25)	938	2.13	43.5	10.6

4.3.1.10 Comparison with Numerical and Empirical Models

FEM simulation results for $d/s = 1$ and 4 are compared to mathematical models of for linear elastic behavior under uniaxial tension. Figure 4.30 shows the stress applied normal to the infinite number of holes [43] create maximum stress concentration around the pores: $\sigma_{max\_theo} = 3230$ MPa for $d = s = 100$ μm which is slightly lower than the simulation result of $\sigma_{max\_FEM} \approx 2950$ MPa. Similarly, $d/s = 4$ ratio sample has $\sigma_{max\_theo} > 5000$ MPa is slightly higher than the FEM results $\sigma_{max\_FEM} \approx 4900$ MPa. This is because AB-stacking morphology contributes slightly to the overall deformation mechanism by reducing stresses perpendicular to the loading direction. This result is also confirmed by the complex stress-state analysis done by [80], where angle of orientation and distance between the pores are taken into consideration.

We have also compared the experimental results of our tensile heterostructures with the generalized empirical and theoretical models using the formulas given in [81]:

$$E^*_{emp} = Ee^{(b*P)} \tag{4.14}$$

$$E^*_{theo} = \frac{E(1-P)^2}{(1+K_m P)} \tag{4.15}$$

Where $E = 94.0$ GPa, $K_m = 3 - 2v$. The empirical constant b was determined to be 3.6 by taking the production methods into account (casting and thermoplastic forming). Figure 4.31 shows a similar exponential decay of the models as the volume fraction porosity (P) increases, which confirms our experimental findings for a wide porosity range of MG heterostructures.

4.3.1.11 Effect of Isothermal Annealing on Mechanical Properties

In addition to the study on the effect of pore spacing on the MG heterostructures' performance, we also varied the properties of the matrix material, and thereby, its plastic zone size. An effective strategy to vary the plastic zone size is to structurally

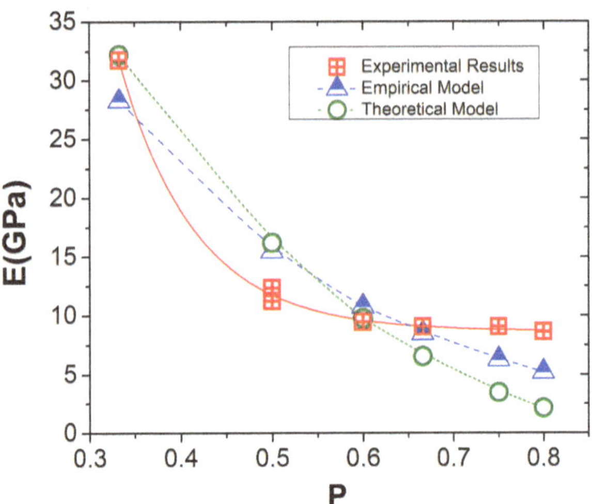

Fig. 4.31 Experimental results compared to empirical and theoretical models. Our experimentally found elastic modulus values fit into exponential decay curve as porosity of the MG heterostructure increases, which is in-line with the calculated values. It is noteworthy to indicate that stress values were normalized with respect to the overall width of the monolithic sample

relax the MG through thermal annealing [39], where structural relaxation (embrittlement) can be studied in a wide range of MG alloys.

Tensile samples are fabricated out of $Zr_{35}Ti_{30}Cu_{7.5}Be_{27.5}$ MG former with $d/s = 1.5$, which were annealed at 558 K, ~25 K below its calorimetric glass transition temperature. The degree of structural relaxation is determined through enthalpy recovery [39, 41, 68, 69, 82–86], and its effect on mechanical behavior through bending tests (Fig. 4.32b). The crystallization was achieved by heating the MG heterostructure to 420 °C (above-T_g) for 1 h. Differential scanning calorimeter (DSC) and X-ray diffraction (XRD) measurements were conducted to determine the effect of different annealing processes on thermal and structural properties. Tensile characterization shows that strain to failure and U_T are dramatically affected by annealing (decrease in R_p) and crystallization (Fig. 4.32a, b)). The completely structurally relaxed MG's for 36 h and longer, as well as the crystallized MG former approximate ideal brittle behavior with $K_{1C} < 5$ MPalm [87], and exhibit a plastic zone size below 1 μm. In such MG heterostructures $R_p < s$, and as a consequence, the overall strain decreases to the elastic limit of approximately 2 %. Even though K_{1C} is not directly quantified in this work, it can be concluded from Fig. 4.32b that for the as cast and 28 h annealed heterostructures $R_p > s$, hence showing large tensile ductility [43].

4.3.2 *Investigation of MG Composites Using FEM Analysis*

In this chapter, it has been shown that the properties of MG heterostructures can be manipulated by tailoring the properties of the pores as the second phase. To understand the influence of the second phase material in a MG heterostructure of AB-stacking morphology, we also conducted FEM simulations with $Zr_{35}Ti_{30}Cu_{7.5}Be_{27.5}$ MG including different second phases ($G_{Graphite}$ ~3.8 GPa, G_{Al} ~26.3 GPa, G_{Ni} ~76.3 GPa) ($d/s = 1$) comparatively with MG heterostructures ($Zr_{35}Ti_{30}Cu_{7.5}Be_{27.5}$

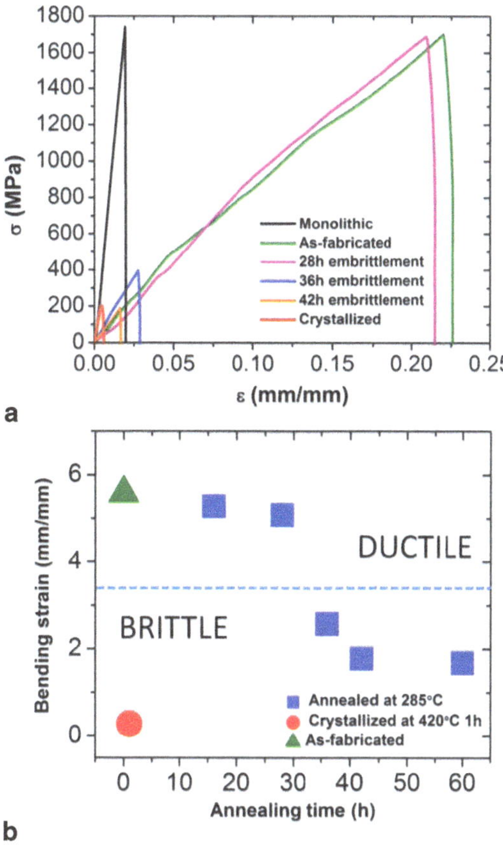

Fig. 4.32 Influence of sub-T_g annealing and crystallization on the ductility of hetero-structures with $d/s = 1.5$. **a** For samples that are completely structurally relaxed at 558 K, as well as for crystallized samples, ductility decreased dramatically. **b** Bending test reveal a significant decrease in plasticity for annealing times exceeding 28 h at 558 K [43]

MG and pores) with $G_{Matrix(Zr\text{-}MG)} = 31.7$ GPa in linear elastic mode. von Mises stress distribution (Fig. 4.33) indicates that when the shear modulus (similarly the elastic modulus) of the second phase is smaller than the matrix MG, which is the case for graphite and aluminum, part of the stress can be transferred to the second phase. This results in less stress percolations in the matrix along the pores perpendicular to the deformation axis. However, when the elastic modulus of the second phase is much higher than the MG, higher stress concentrations are observed within the second phase, as well as in the MG matrix between the pores parallel to the loading direction.

Similarly, the evaluation between the shear stress values show that the best compromise is obtained when the Elastic modulus of the second phase is relatively lower than the matrix MG, i.e., aluminum (Fig. 4.34). Shear stress between the diagonal second phase features is an order of magnitude smaller than for the hetero-structures with pores, graphite, and Ni. Very high stress accumulations are present on the 45° angle for the second phases except Al, causing the material to yield at lower stresses.

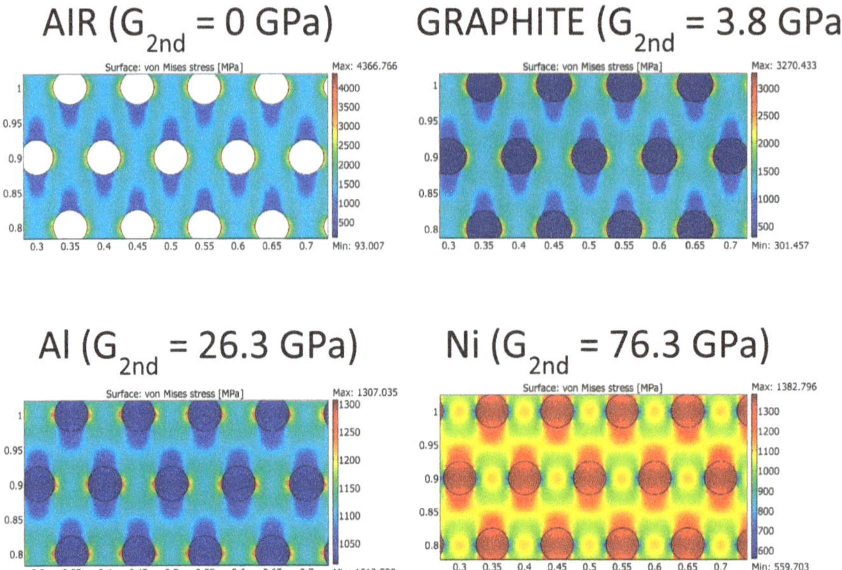

Fig. 4.33 von Mises stress distribution of the AB-stacking heterostructures with a second phase of different shear modulus. A second phase with significantly different shear modulus than the matrix results in very high stress concentrations

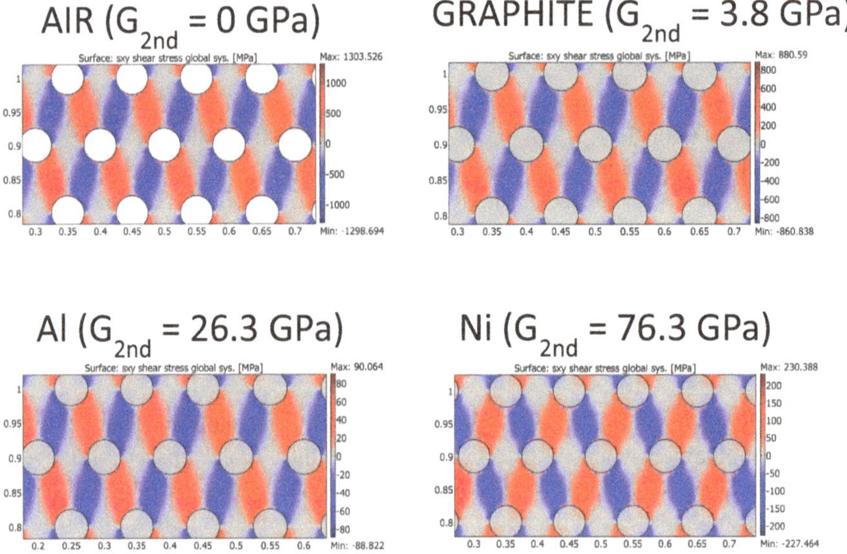

Fig. 4.34 Shear stress comparison of heterostructures with different second phases. The best compromise has been achieved when the shear modulus of the second phase is relatively lower than the matrix MG

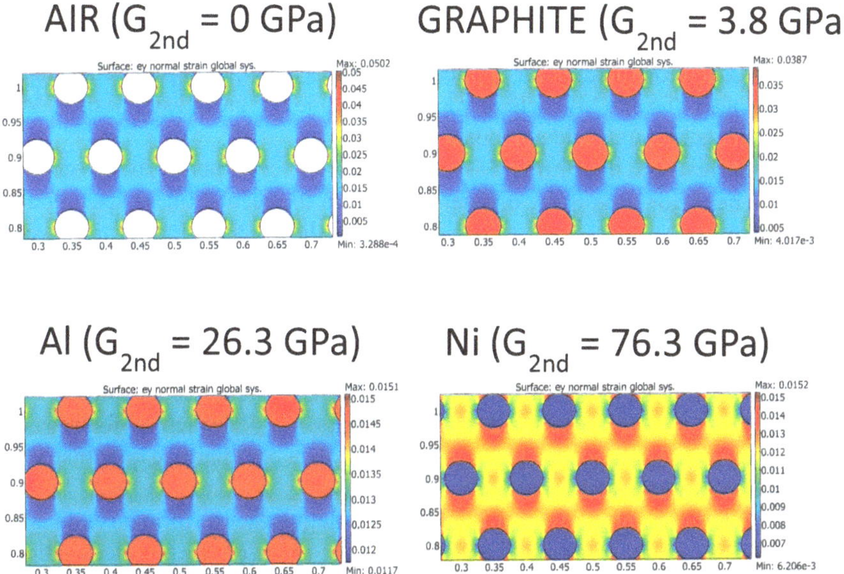

Fig. 4.35 Amount of elasticity as a function of normal strain in the loading direction. The highest elastic strain (~5 %) is obtained for $G_{2ndphase} \approx 0$ (pores)

Strain values in *y*-axis shown in Fig. 4.35 indicates that the heterostructure with pores are showing maximum of 5 % elasticity in comparison with the sum of the total strain in the second phase and MG matrix compared to heterostructures with graphite (3.9 %), Al (3 %), and Ni (3 %) second phases.

To summarize, the softer second phase additions dissipates the stress more homogenously throughout the material and create less stress percolations. On the other hand, in real experiments, the soft and hard second phase additions (except pores) of the heterostructures yields at much lower strains, which would cause the MG matrix phase to carry the stress only by itself after reaching to σ_y of the second phase. According to these findings, the heterostructures with pores is predicted to be a very good combination of elastic strain and the stress distribution around the pores, but can further be enhanced by an addition of a second phase with high elasticity and fracture strength together with a slightly lower shear modulus (e.g., MG–MG composites).

4.3.3 General Findings and Conclusions

The aim of our artificial microstructure study has been to identify and understand the mechanisms that result in tensile ductility in MG heterostructures such as foams, cellular heterostructures, and to a limited extent, composites. The strength and purpose of our method is the ability to precisely vary individual microstructural features completely independent, and determine the effect of such isolated variation on the mechanical response. As such, our method is not intended as a scalable method to fabricate MG heterostructures for applications, but as a method to help understand the effectiveness of some MG heterostructures to guide design and synthesis of MG foams, cellular structures, and composites.

Our results revealed three critical aspects of a MG heterostructure, which control mechanical behavior. One is the difference between the second phase spacing and the plastic zone size of the MG, $s\text{-}R_p$. For $s\text{-}R_p<0$, shear bands that form between pores do not grow to a length where they transform into a crack. As a consequence, global tensile ductility is achieved through the formation of multiple shear bands. For $s\text{-}R_p>0$, a shear band can develop into a crack prior to strain energy release into the pore and consequently, only one or a few shear bands form prior to crack formation, which impairs tensile ductility. Our finding is broadly confirmed in MG heterostructures with a soft second phase, i.e., $G_{2nd}<G_{matrix}$, such as foams, cellular structures, and composites [43]; however, till date could not be quantitatively proven due to experimental limitations.

The $s\text{-}R_p$ criterion alone is insufficient in predicting the performances of MG heterostructures with a soft second phase due to the effect of stress concentrations on the deformation mechanism. In addition to $s\text{-}R_p$, we identified the ratio of second phase size to spacing as the second criterion for toughness and tensile ductility optimization. Maximum toughness is achieved for $d/s\approx 1$, which corresponds to a density of ~50 %. This value for d/s exhibits the best compromise between enhanced stress concentrations which limits plastic deformation, and thin interporous regions, which enhance plasticity in cellular structures.

Even though our design criterion can be applied to cellular MGs, the applicability for the effective design of MG composites is limited, which is due to the complex nature of the composites. Experimental findings suggest the influences of shape and volume fraction of the second phase, interfacial strength and toughness, softness of the second phase reflected in shear modulus, and even dislocation and twinning in the second phase. Our finding of the drastically different mechanical response between MG heterostructures with AB- and AA-stacking exemplifies the importance of the second phase morphology, and the overall complex nature of the toughening mechanism in MG composites.

We argue that the effectiveness of MG heterostructures also depends on the size (cross-section width perpendicular to the loading direction) of the sample, w, in comparison to s. Such a size effect originates from the fact that shear bands do not support tensile forces during the actual shearing event. Since the cross-sectional area supporting a tensile load in samples, where w is not much larger than s, is significantly reduced when one shear band forms, the applied load results in exceeding σ_y throughout the sample. Therefore, small samples with w comparable to s fail catastrophically along one shear band. In large samples, $w>>s$ (assuming $s\text{-}R_p<0$), upon formation of a shear band between two neighboring pores, stresses between most of the other pores typically do not exceed σ_y since the stress increase with the area reduction, $w\text{–}s$, due to the initial shearing event is small. This finding also explains why all reported MG composites exhibiting tensile ductility obey $w>>s$.

Our results also reveal the possibility to design elasticity through the elasticity and the modulus of the second phase in MG heterostructures. We have demonstrated up to 6.0 % of elastic strain for the heterostructure with $d = 100$ μm and $d/s = 2$, and an AB-stacking order. Morphology in MG foams and composites is AB-like; however, in MG foams and composites, an increase in elasticity compared to the monolithic MG has not been reported yet. For MG composites, this is due to the properties of the second phase, which exhibits low elasticity (typically below 0.5 %, and comparable modulus to the MG matrix), thereby preventing increased heterostructure elasticity. MG foams with controlled features have not been tested in tension yet.

In summary, we identified the most effective porous heterostructure to create tensile ductility and toughness in MGs by featuring a spacing of the second phase, which coincides with the plastic zone size of the MG and with the second phase size. In addition, we found that heterostructure morphology is crucial and the size of the sample should be large compared to the second phases' spacing.

References

1. Conner RD, Johnson WL, Paton NE, Nix WD. Shear bands and cracking of metallic glass plates in bending. J Appl Phys. 2003;94:904.
2. Wada T, Inoue A, Greer AL. Enhancement of room-temperature plasticity in a bulk metallic glass by finely dispersed porosity. Appl Phys Lett. 2005;86:251907-1.
3. Hofmann DC, Suh JY, Wiest A, Duan G, Lind ML, Demetriou MD, Johnson WL. Designing metallic glass matrix composites with high toughness and tensile ductility. Nature. 2008;451:1085.
4. Volkert CA, Donohue A, Spaepen F. Effect of sample size on deformation in amorphous metals. J Appl Phys. 2008;103:083539-1.
5. Kumar G, Desai A, Schroers J. Bulk metallic glass: the smaller the better. Adv Mater. 2011;23:461.
6. Argon AS. Plastic-deformation in metallic glasses. Acta Metall. 1979;27:47.
7. Schafer J, Blumenstein C, Meyer S, Wisniewski M, Claessen R. New model system for a one-dimensional electron liquid: self-organized atomic gold chains on Ge (001). Phys Rev Lett. 2008;101:236802-1.
8. Xing LQ, Li Y, Ramesh KT, Li J, Hufnagel TC. Enhanced plastic strain in Zr-based bulk amorphous alloys. Phys Rev B. 2001;64:180201-1.
9. Schroers J, Johnson WL. Ductile bulk metallic glass. Phys Rev Lett. 2004;93:255506-1.
10. Das J, Tang MB, Kim KB, Theissmann R, Baier F, Wang WH, Eckert J. "Work-hardenable" ductile bulk metallic glass. Phys Rev Lett. 2005;94:205501-1.
11. Liu YH, Wang G, Wang RJ, Zhao DQ, Pan MX, Wang WH. Super plastic bulk metallic glasses at room temperature. Science. 2007;315:1385:245502-1.
12. Chen MW, Inoue A, Zhang W, Sakurai T. Extraordinary plasticity of ductile bulk metallic glasses. Phys Rev Lett. 2006;96:245502-1.
13. Lee SW, Huh MY, Fleury E, Lee JC. Crystallization-induced plasticity of Cu-Zr containing bulk amorphous alloys. Acta Mater. 2006;54:349.
14. Kumar G, Ohkubo T, Mukai T, Hono K. Plasticity and microstructure of Zr-Cu-Al bulk metallic glasses. Scripta Mater. 2007;57:173.
15. Sarac B, Schroers J. From brittle to ductile: Density optimization for Zr-BMG cellular structures. Scripta Mater. 2013;68:921.

16. Fan C, Inoue A. Ductility of bulk nanocrystalline composites and metallic glasses at room temperature. Appl Phys Lett. 2000;77:46.
17. Szuecs F, Kim CP, Johnson WL. Mechanical properties of Zr56.2Ti13.8Nb5.0Cu6.9Ni5.6 Be12.5 ductile phase reinforced bulk metallic glass composite. Acta Mater. 2001;49:1507.
18. He G, Eckert J, Loser W. Stability, phase transformation and deformation behavior of Ti-base metallic glass and composites. Acta Mater. 2003;51:1621.
19. Ott RT, Fan C, Li J, Hufnagel TC. Structure and properties of Zr-Ta-Cu-Ni-Al bulk metallic glasses and metallic glass matrix composites. J Non-Cryst Solids. 2003;317:158.
20. Lee ML, Li Y, Schuh CA. Effect of a controlled volume fraction of dendritic phases on tensile and compressive ductility in La-based metallic glass. Acta Mater. 2004;52:4121.
21. Inoue A, Zhang W, Tsurui T, Yavari AR, Greer AL. Unusual room-temperature compressive plasticity in nanocrystal-toughened bulk copper-zirconium glass. Phil Mag Lett. 2005;85:221.
22. Hajlaoui K, Doisneau B, Yavari AR, Botta WJ, Zhang W, Vaughan G, Kvick A, Greer AL. Unusual room temperature ductility of glassy copper-zirconium caused by nanoparticle dispersions that grow during shear. Mater Sci Eng a-Struct. 2007;449:105.
23. Eckert J, Das J, Pauly S, Duhamel C. Mechanical properties of bulk metallic glasses and composites. J Mater Res. 2007;22:285.
24. Evans AG, Hutchinson JW, Fleck NA, Ashby MF, Wadley HNG. The topological design of multifunctional cellular metals. Prog Mater Sci. 2001;46:309.
25. Demetriou MD, Veazey C, Harmon JS, Schramm JP, Johnson WL. Stochastic metallic-glass cellular structures exhibiting benchmark strength. Phys Rev Lett. 2008;101:145702-1.
26. Bouwhuis BA, Hibbard GD. Relative significance of in-situ work-hardening in deformation-formed micro-truss materials. Mater Sci Eng a-Struct. 2010;527:565.
27. Kirkland NT, Kolbeinsson I, Woodfield T, Dias G, Staiger MP. Processing-property relationships of as-cast magnesium foams with controllable architecture. Int J Mod Phys B. 2009;23:1002.
28. Schroers J, Veazey C, Johnson WL. Amorphous metallic foam. Appl Phys Lett. 2003;82:370.
29. Brothers AH, Dunand DC. Syntactic bulk metallic glass foam. Appl Phys Lett. 2004;84:1108.
30. Schroers J, Veazey C, Demetriou MD, Johnson WL. Synthesis method for amorphous metallic foam. J Appl Phys. 2004;96:7723.
31. Brothers AH, Dunand DC. Ductile bulk metallic glass foams. Adv Mater. 2005;17:484.
32. Brothers AH, Dunand DC, Zheng Q, Xu J. Amorphous Mg-based metal foams with ductile hollow spheres. J Appl Phys. 2007;102:023508-1.
33. Schuh CA, Hufnagel TC, Ramamurty U. Overview No. 144—mechanical behavior of amorphous alloys. Acta Mater. 2007;55:4067.
34. Greer AL. Metallic glasses ... on the threshold. Mater Today. 2009;12:14.
35. Hays CC, Kim CP, Johnson WL. Microstructure controlled shear band pattern formation and enhanced plasticity of bulk metallic glasses containing in situ formed ductile phase dendrite dispersions. Phys Rev Lett. 2000;84:2901.
36. Fan C, Ott RT, Hufnagel TC. Metallic glass matrix composite with precipitated ductile reinforcement. Appl Phys Lett. 2002;81:1020.
37. He G, Eckert J, Loser W, Schultz L. Novel Ti-base nanostructure-dendrite composite with enhanced plasticity. Nat Mater. 2003;2:33.
38. Guo H, Yan PF, Wang YB, Tan J, Zhang ZF, Sui ML, Ma E. Tensile ductility and necking of metallic glass. Nat Mater. 2007;6:735.
39. Kumar G, Rector D, Conner RD, Schroers J. Embrittlement of Zr-based bulk metallic glasses. Acta Mater. 2009;57:3572.
40. Jang DC, Greer JR. Transition from a strong-yet-brittle to a stronger-and-ductile state by size reduction of metallic glasses. Nat Mater. 2010;9:215.
41. Kumar G, Prades-Rodel S, Blatter A, Schroers J. Unusual brittle behavior of Pd-based bulk metallic glass. Scripta Mater. 2011;65:585.
42. Sarac B, Ketkaew J, Popnoe DO, Schroers J. Honeycomb structures of bulk metallic glasses. Adv Funct Mater. 2012;22:3161.

43. Sarac B, Schroers J. Designing tensile ductility in metallic glasses. Nat Commun. 2013;4:2158-1.
44. Schroers J, Pham Q, Desai A. Thermoplastic forming of bulk metallic glass—a technology for MEMS and microstructure fabrication. J Microelectromech S. 2007;16:240.
45. Schroers J. Processing of bulk metallic glass. Adv Mater. 2010;22:1566.
46. Sarac B, Kumar G, Hodges T, Ding SY, Desai A, Schroers J. Three-dimensional shell fabrication using blow molding of bulk metallic glass. J Microelectromech Syst. 2011;20:28.
47. Schroers J, Hodges TM, Kumar G, Raman H, Barnes AJ, Quoc P, Waniuk TA. Thermoplastic blow molding of metals. Mater Today. 2011;14:14.
48. Gibson LJ, Ashby MF, Schajer GS, Robertson CI. The mechanics of two-dimensional cellular materials. Proc Roy Soc Lond a Mat. 1982;382:25.
49. Ashby MF. The mechanical-properties of cellular solids. Metal Trans A. 1983;14:1755.
50. Fleck NA, Chen C, Lu TJ. Effect of imperfections on the yielding of two-dimensional foams. J Mech Phys Solids. 1999;47:2235.
51. Papka SD, Kyriakides S. Inplane compressive response and crushing of honeycomb. J Mech Phys Solids. 1994;42:1499.
52. Saotome Y, Imai K, Shioda S, Shimizu S, Zhang T, Inoue A. The micro-nanoformability of Pt-based metallic glass and the nanoforming of three-dimensional structures. Intermetallics. 2002;10:1241.
53. Henann DL, Srivastava V, Taylor HK, Hale MR, Hardt DE, Anand L. Metallic glasses: viable tool materials for the production of surface microstructures in amorphous polymers by micro-hot-embossing. J Micromech Microeng. 2009;19:115030-1.
54. Takenaka K, Saidoh N, Nishiyama N, Inoue A. Fabrication and nano-imprintabilities of Zr-, Pd- and Cu-based glassy alloy thin films. Nanotechnology. 2011;22:105302-1.
55. Ohsaka K, Chung SK, Rhim WK, Peker A, Scruggs D, Johnson WL. Specific volumes of the Zr41.2Ti13.8Cu12.5Ni10.0Be22.5 alloy in the liquid, glass, and crystalline states. Appl Phys Lett. 1997;70:726.
56. Roberts RB. Thermal-expansion reference data—silicon 300-850-K. J Phys D Appl Phys. 1981;14:L163.
57. Gibson LJ, Ashby MF. Cellular solids—structure and properties. 2001.
58. Superplastic Aluminum Alloys. Total Materia. http://www.keytometals.com/page.aspx?ID=CheckArticle&site=ktn&NM=264. 2011.
59. Low-density Polyethylene. Matbase—Material Properties Database. http://www.matbase.com/material-categories/natural-and-synthetic-polymers/commodity-polymers/material-properties-of-low-density-polyethylene-ldpe.html. Accessed 1 Sept 2014.
60. Duan G, Wiest A, Lind ML, Li J, Rhim WK, Johnson WL. Bulk metallic glass with benchmark thermoplastic processability. Adv Mater. 2007;19:4272.
61. Trexler MM, Thadhani NN. Mechanical properties of bulk metallic glasses. Prog Mater Sci. 2010;55:759.
62. Zhang ZF, Eckert J, Schultz L. Difference in compressive and tensile fracture mechanisms of Zr59CU20Al10Ni8Ti3 bulk metallic glass. Acta Mater. 2003;51:1167.
63. Zinc. Canada Metal Limited. http://www.canadazincmetals.com. Accessed 1 Sept 2014.
64. Gilbert CJ, Ritchie RO, Johnson WL. Fracture toughness and fatigue-crack propagation in a Zr-Ti-Ni-Cu-Be bulk metallic glass. Appl Phys Lett. 1997;71:476.
65. Nagel C, Ratzke K, Schmidtke E, Faupel F. Positron-annihilation studies of free-volume changes in the bulk metallic glass Zr65Al7.5Ni10Cu17.5 during structural relaxation and at the glass transition. Phys Rev B. 1999;60:9212.
66. Slipenyuk A, Eckert J. Correlation between enthalpy change and free volume reduction during structural relaxation of Zr55Cu30Al10Ni5 metallic glass. Scripta Mater. 2004;50:39.
67. Kanungo BP, Glade SC, Asoka-Kumar P, Flores KM. Characterization of free volume changes associated with shear band formation in Zr- and Cu-based bulk metallic glasses. Intermetallics. 2004;12:1073.
68. Murali P, Ramamurty U. Embrittlement of a bulk metallic glass due to sub-T-g annealing. Acta Mater. 2005;53:1467.

69. Lewandowski JJ. Effects of annealing and changes in stress state on fracture toughness of bulk metallic glass. Mater Trans. 2001;42:633.
70. Wang AJ, McDowell DL. Effects of defects on in-plane properties of periodic metal honeycombs. Int J Mech Sci. 2003;45:1799.
71. Papka SD, Kyriakides S. Experiments and full-scale numerical simulations of in-plane crushing of a honeycomb. Acta Mater. 1998;46:2765.
72. Conner RD, Dandliker RB, Johnson WL. Mechanical properties of tungsten and steel fiber reinforced Zr41.25Ti13.75Cu12.5Ni10Be22.5 metallic glass matrix composites. Acta Mater. 1998;46:6089.
73. Park BJ, Chang HJ, Kim DH, Kim WT, Chattopadhyay K, Abinandanan TA, Bhattacharyya S. Phase separating bulk metallic glass: a hierarchical composite. Phys Rev Lett. 2006;96:245503-1.
74. Hofmann DC, Suh JY, Wiest A, Lind ML, Demetriou MD, Johnson WL. Development of tough, low-density titanium-based bulk metallic glass matrix composites with tensile ductility. Proc Natl Acad Sci U S A. 2008;105:20136.
75. Chen G, Cheng JL, Liu CT. Large-sized Zr-based bulk-metallic-glass composite with enhanced tensile properties. Intermetallics. 2012;28:25.
76. Liu ZQ, Li R, Liu G, Su WH, Wang H, Li Y, Shi MJ, Luo XK, Wu GJ, Zhang T. Microstructural tailoring and improvement of mechanical properties in CuZr-based bulk metallic glass composites. Acta Mater. 2012;60:3128.
77. Yi J, Xia XX, Zhao DQ, Pan MX, Bai HY, Wang WH. Micro-and nanoscale metallic glassy fibers. Adv Eng Mater. 2010;12:1117.
78. Myers MA. Mechnical metallurgy: principles and applications. New Jersey: Prentice Hall; 1984.
79. Qu RT, Calin M, Eckert J, Zhang ZF. Metallic glasses: notch-insensitive materials. Scripta Mater. 2012;66:733.
80. Bidhar S, Kuwazuru O, Hangai Y, Yano T, Utsunomiya T, Yoshikawa N. Empirical prediction of stress concentration factor for a pair of spherical cavities. IEEE. 2011;1.
81. Herakovich CT, Baxter SC. Influence of pore geometry on the effective response of porous media. J Mater Sci. 1999;34:1595.
82. Chen HS. Thermal and mechanical stability of metallic glass ferromagnets. Scripta Metal. 1977;11:367.
83. Wu TW, Spaepen F. Embrittlement of metallic glasses. J Metal. 1984;36:54.
84. Wu TW, Spaepen F. The relation between embrittlement and structural relaxation of an amorphous metal. Philos Mag B. 1990;61:739.
85. Lewandowski JJ, Wang WH, Greer AL. Intrinsic plasticity or brittleness of metallic glasses. Phil Mag Lett. 2005;85:77.
86. Ramamurty U, Lee ML, Basu J, Li Y. Embrittlement of a bulk metallic glass due to low-temperature annealing. Scripta Mater. 2002;47:107.
87. Waniuk TA, Busch R, Masuhr A, Johnson WL. Equilibrium viscosity of the Zr41.2Ti13.8Cu12.5Ni10Be22.5 bulk metallic glass-forming liquid and viscous flow during relaxation, phase separation, and primary crystallization. Acta Mater. 1998;46:5229.

Chapter 5
General Conclusions and Outlook

5.1 General Conclusions

Within this thesis, we introduced a novel approach called artificial microstructures to determine microstructure–property relationships in complex materials. The biggest advantage of this technique over other conventional methods is the systematic data analysis and quantification of results, where we determine the effect of each feature on mechanical properties through completely independent feature variation. We utilized this approach to address two important problems in metallic glasses (MGs). The first problem was to understand the behavior of MGs in hexagonal cellular structures. Here, we found that the deformation can be controlled and manipulated by changing the relative density. As a consequence, three major deformation regions are discovered: collective buckling showing nonlinear elasticity, localized failure exhibiting a brittle-like deformation, and global sudden failure with negligible plasticity. The ideal density for optimal mechanical properties was determined to be ~25.0 %, which is within the local failure deformation regime. Enhancement in mechanical properties in MG cellular structures was achieved by stress optimization through corner-fillets, which doubled the strength at the expense of 0.2 % density increase. Besides, energy absorption of MG cellular structures exceeds cellular structures of most other materials due to the utilization of a size effect.

The second problem was to create a toughening mechanism in MG tensile heterostructures. An important design criterion is that the second phase spacing should be larger than the plastic zone size of the MG, which was quantitatively determined by the experimental results for the first time in this study. Another important criterion was determined to be the sample size, which should be large compared to the second phases' spacing ($w >> s$). We revealed the pores' ratio of diameter to spacing (d/s) is one of the other major factors in controlling the mechanical properties of MG heterostructures, where the optimum strength/strain values were obtained when d/s value is set to ~1. Our results also provide a guidance to design elasticity through second phase morphology (AB pore stacking) in MG heterostructures, which depends on the second phase elasticity and the shear modulus.

B. Sarac, *Microstructure-Property Optimization in Metallic Glasses*, Springer Theses,
DOI 10.1007/978-3-319-13033-0_5

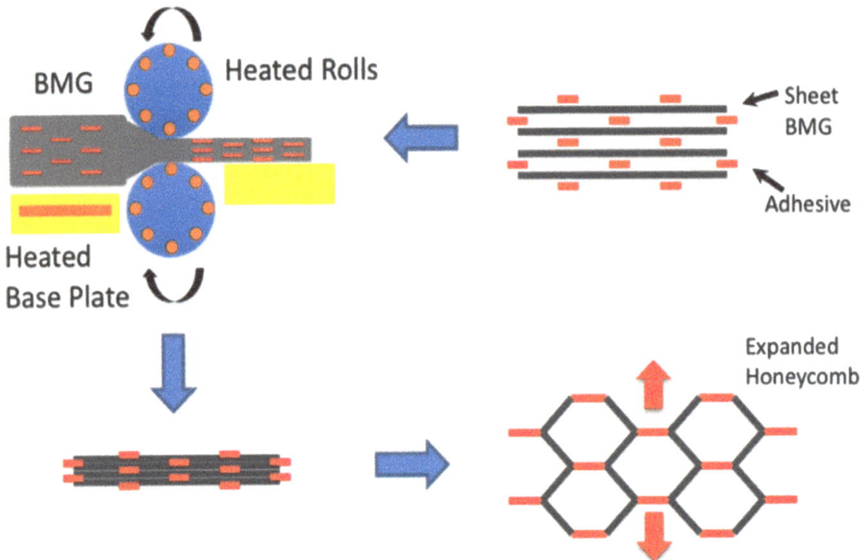

Fig. 5.1 Schematics of 3D cellular structure fabrication by uniaxial expansion of the cells

5.2 Push the Limit: 3D Metallic Glass Structures

Two-dimensional artificial microstructure approach provides versatility and predictability to analyze the microstructure–property relationships, where the first systematic study about the MG heterostructures has been presented in this thesis. However, this approach can only draw conclusions about the base layer of 3D cellular structures, which is the 2D cut. It has to be determined how well the 2D cuts represent the behavior of 3D structures. To address this question, our current objective is to fabricate 3D MG structures by combining multiple processes such as joining, compression molding, and blow molding of metallic glasses.

Figure 5.1 shows the fabrication steps of expanded cellular material production. MG sheets are produced by heating the MG cast rods to the processing temperature, at which they become malleable and can be rolled to a sheet form. BMG sheets are glued together using an adhesive (i.e., hot cure epoxy), where the glued section will act as joints once we form 3D cellular structures. The sheets will then be hot rolled to establish strong adhesion of these joints. The cellular structure is consequently shaped by stretching the deformed shape from the joint sections until it forms the periodic 3D structure.

Another technique considered is to utilize direct imprinting method. The features patterned on the rollers can easily be transferred as the MG sheet is rolled at its $T_{process}$. Figure 5.2a shows the repetitive pattern formed using the TPF-based hot rolling of the MG. The patterned MG sheets can further be bonded together using epoxy cure or reaction bonding (Fig. 5.2b).

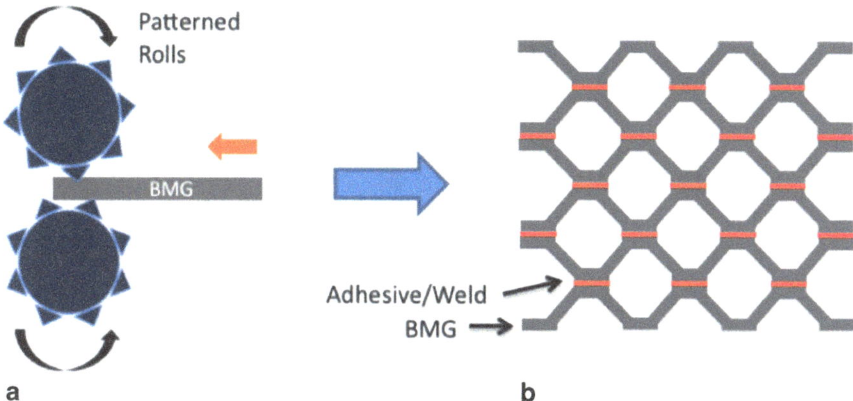

Fig. 5.2 **a** Transfer of the desired features through direct imprinting. Metallic glass (MG) sheet is forced into the rotating mills with the embedded patterns. **b** The thermoplastically formed sheets can be subsequently bonded to each other using an adhesive or a welding method

5.3 Multiple Material Artificial Microstructures

Within this thesis, only one material with pores as second phases was considered so far. Our new target is to create more versatility by introducing a second phase. The challenge to add a different second phase with this adopted technique is that a relatively weak interface between the matrix and the second phase is very likely to form due to the difference in thermal expansion coefficients, which results in relatively different linear shrinkage when cooled down to room temperature:

$$\frac{\Delta L}{L_0} = \alpha \Delta T \tag{5.1}$$

with $\Delta L/L_0$= Linear Shrinkage, α= Thermal expansion coefficient (°C^{-1}), ΔT= Temperature difference between the processing temperature of the second phase and the room temperature.

We are now proposing a new method to create MG composites with the systematically arranged second phase structures. To estimate the stress and strain distribution, finite element simulations were conducted with different second phase materials. Section 4.3.2 indicates that stress concentration can be more homogenously distributed throughout the structure especially when the second phase has slightly lower shear modulus. The design consists of pores with AB-stacking etched into the Si mold through photolithography and DRIE (Fig. 5.3). The fabrication is in twofold: first, the base layer with the pillars standing on top is fabricated out of second phase material using TPF-based compression molding (Fig. 5.3a). The Si mold is subsequently etched out using KOH. The second step consists of compression molding of a matrix MG, which fills the gaps between the pillars (Fig. 5.3b, c). Thereby, the glassy matrix contracts the pillars of the second phase (the second

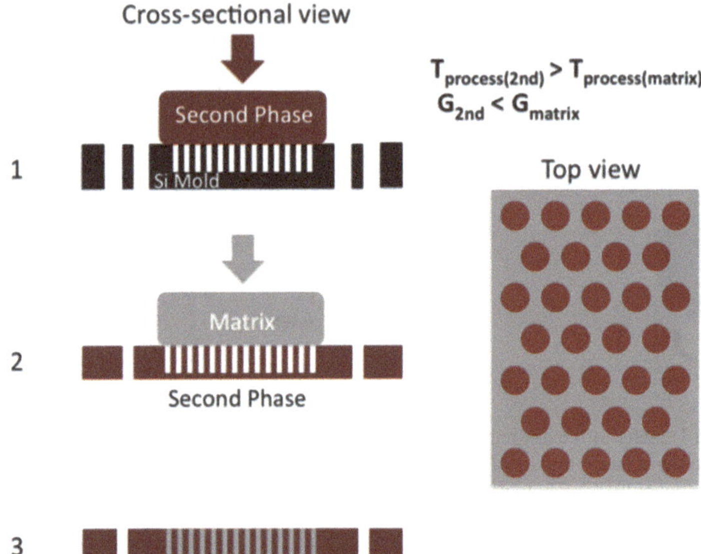

Fig. 5.3 Fabrication steps of two-phase composite sample out of different metallic glasses (MGs). The first metallic glass replicates the embedded features of the Si mold by TPF. The silicon mold is etched out, and the matrix material is filled into the cavities of the second phase using TPF at a relatively lower processing temperature. The cross-sectional view of the final shape shows the AB-stacking morphology of the second phase, where the second phase features can be precisely controlled

phase remains solid through TPF) while cooling, resulting in a strong mechanical interlocking system. In order to retain the fully amorphous state of the second phase and have a relatively stiffer matrix phase, the matrix material should have a lower processing temperature and a higher modulus compared to the second phase. The possible candidates for the glassy matrix phase are $Zr_{35}Ti_{30}Cu_{7.5}Be_{27.5}$ (G=31.7 GPa [1], $T_{process}$~420°C), $Pt_{57.5}Cu_{14.7}Ni_{5.3}P_{22.5}$ (G=33.3 GPa [2], $T_{process}$=260°C), $Pd_{43}Cu_{27}Ni_{10}P_{20}$ (G=33.4 GPa [3], $T_{process}$=310°C), whereas for the second phase a ductile crystalline metal such as superplastic aluminum Alloys (e.g., SUPRAL 100, G=27.7 GPa [4], $T_{process}$=470°C) or tin (G=18.0 GPa [5], $T_{process}$ =320°C) can be selected in accordance with the matrix material. Alternatively, two different MG alloys with different linear expansion coefficients can also be utilized. The ex situ composite sample, where the second phase features are tailorable by CAD design, will be characterized under quasi-static uniaxial tension, where the stress–strain curve is going to provide information for design improvements for optimum mechanical properties in composite structures. Incorporation of this technique into microstructural-property investigation is the preliminary goal to understand the mechanical behavior of these composite structures under tension.

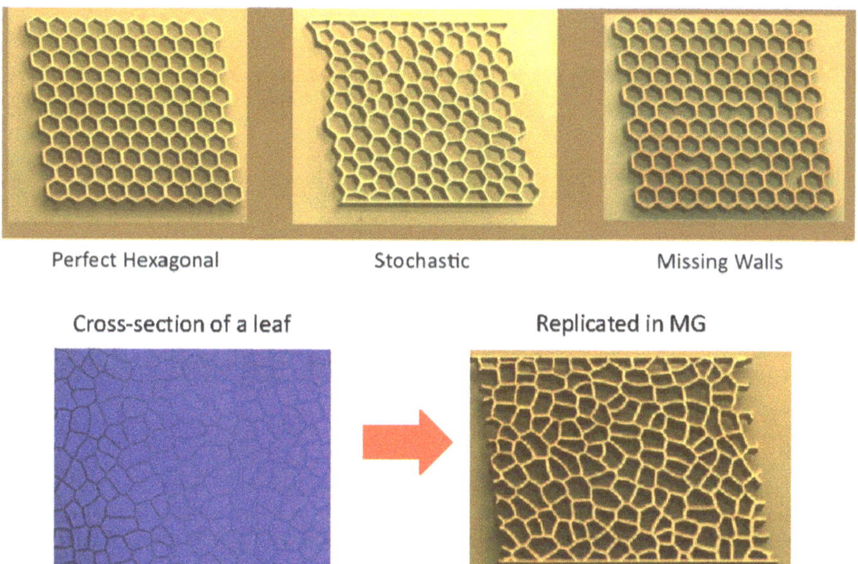

Fig. 5.4 Sample microstructures with imperfections and randomness compared to perfect hexagonal honeycombs (*top*), as well as replication of structures using image conversion. Wide range of geometrical versatility and intricacy is possible with photolithography and TPF-based compression molding

5.4 Nonperiodic Cellular Structures and Flaw Tolerance

Understanding the microstructure–property relationship is intimately connected to understanding the effect of defects or imperfections on mechanical properties. The proposed method is suited to the assessment of these effects, since it allows one to create both individual defects and any combination of defects. Furthermore, imperfections in the microstructure, e.g., size, spacing, shape, and their distribution, can be precisely created. We are currently involved in designing periodic and nonperiodic (stochastic) foams having different imperfections to understand the flaw tolerance in MG cellular structures (Fig. 5.4 top).

We are also interested in biomimicking the natural structures and examine their mechanical properties under different loading conditions. We replicate cross-sections of microstructures such as leafs or insect wings, or 2D spider webs from the real images. We use multiple steps to realize these microstructures out of MGs: Image processing (Photoshop), conversion of the image into a drawing format (Layout Editor) and structural manipulation (AutoCAD), and finally, realization of the structure through photolithography and TPF-based compression molding (Fig. 5.4 bottom).

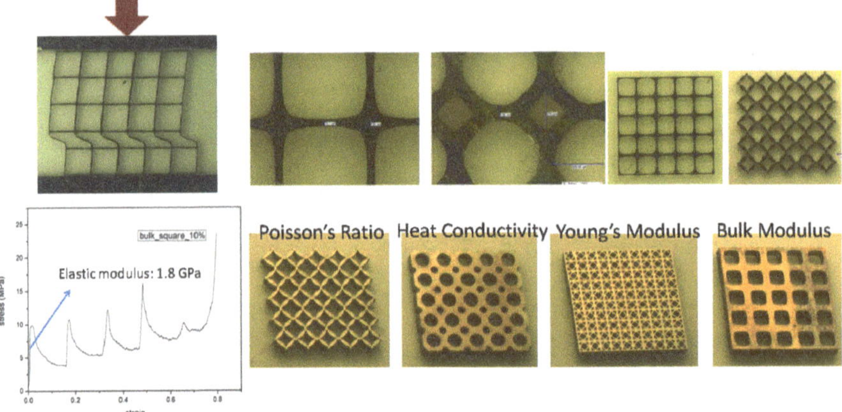

Fig. 5.5 Optimization of cellular structures using computational algorithms. The stress–strain data obtained by in-plane compressive deformation will be used to evaluate the mechanical properties of the related cellular structure, as well as to create complex microstructures optimized for multiple properties

5.5 Algorithmic Topological Optimization

Another subject we are currently interested in is to improve multiple mechanical properties of MGs concomitantly using computational algorithms. Our current target comprises three different designs: Superelastic cellular structures, structures showing negative Poisson's ratio, and structures with strength of a ceramic and plasticity of a polymer. The preliminary designs optimized for different properties, and the illustrative deformation matched with the stress–strain data in Fig. 5.5 will give us guidance for multiple property optimization of MG heterostructures.

5.6 Fracture Toughness in MG Heterostructures

The correlation between spacing of the second phase features and the critical plastic zone size for crack initiation can only be established by precise measurement of the fracture toughness of the MG sample. However, measuring the K_{IC} of the MG is very challenging, and the measured value differs a lot from one to another for the same MG sample. This is because as soon as a crack initiates, it propagates quite rapidly due to no plasticity.

The primary target of this work is to determine K_{IC} values for MG heterostructures with different pore features. For this reason, we proposed the notched heterostructure design shown in Fig. 5.6. The preliminary results show that the deformation of the sample creates shear bands between the pores in the vicinity of a crack, as well as at the tip of the crack, which is an indication of plasticity.

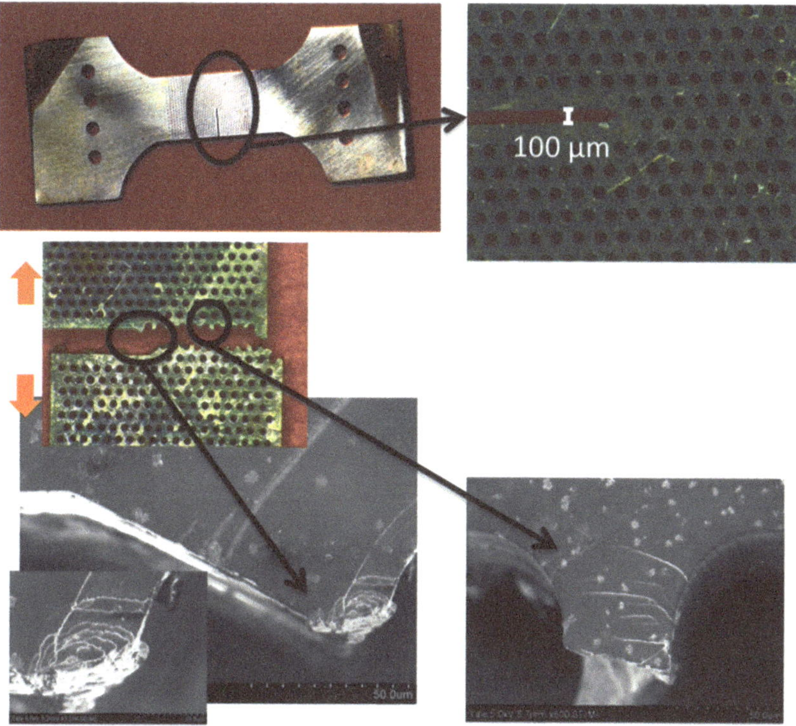

Fig. 5.6 Fracture toughness sample designed for tensile metallic glass (MG) heterostructures. Shear bands between the pores and at the crack tip reveals plasticity

5.7 Other Application Fields of MG Heterostructures

Apart from the techniques mentioned in this chapter, due to MGs' favorable intrinsic properties, 3D microstereolithography (additive manufacturing) techniques can also be adopted to these alloy systems, which can potentially create complicated structures with higher robustness and energy absorption capability with respect to the conventional materials and methods used in this field [6, 7].

Recent studies consisting of modeling of size-dependent behavior of MGs with continuum finite element analysis have garnered remarkable attention due to the semi-empirical formulation of the material model, resulting in accurate prediction of the actual sample deformation. Among these, the material model of Thamburaja [8, 9] can capture the length-scale effects on a nanoscale level and predict the shear localization process at room and elevated temperatures. A gradient-extended energy-based mathematical model presented in the last couple of years describes the shear localization process as a function of free volume generation, shear localization and cohesion, by which multiple length scale MG samples can be analyzed for the infinitesimal [10] and finite strain deformation [11]. Recently, our group has shown that this model can also provide answers for the critical crack length and fracture mechanism of MGs with periodic [12] and stochastic [13] pore distribution. The scalability of this model is conferred by a representative volume element with applied periodic boundary conditions, which creates a possibility to investigate real applications such as beams or foam structures on a macroscale level.

The suggested combined silicon lithography and thermoplastic forming approach as well as other processing routes such as direct imprinting from metallic glass sheets and 3D additive manufacturing within this thesis can serve as a tool to manufacture cellular structures with high functionality for micro-fuel cells or bio-inspired structures. Moreover, the aforementioned methods provide easiness and flexibility in microstructural architecture generation, which can be used for creating porous structures for high shock absorbers, ultraelastic springs or shape-memory alloys for MEMS devices, heat insulation systems and so on.

References

1. Duan G, Wiest A, Lind ML, Li J, Rhim WK, Johnson WL. Bulk metallic glass with benchmark thermoplastic processability. Adv Mater. 2007;19:4272–5.
2. Schroers J, Johnson WL. Ductile bulk metallic glass. Phys Rev Lett. 2004;93:255506-1 (-255506-4).
3. Nishiyama N, Inoue A, Jiang JZ. Elastic properties of Pd40Cu30Ni10P20 bulk glass in supercooled liquid region. Appl Phys Lett. 2001;78:1985–7.
4. Superplastic Aluminum Alloys. Total Materia. 2011 http://www.keytometals.com/page.aspx?ID=CheckArticle&site=ktn&NM=264. Accessed 1 Sept 2011.
5. Tin. Wikipedia, the free encyclopedia. 2014 http://en.wikipedia.org/w/index.php?title=Tin&oldid=640405865. Accessed 1 Sept 2014.
6. Zheng XY, Lee H, Weisgraber TH, Shusteff M, DeOtte J, Duoss EB, Kuntz JD, Biener MM, Ge Q, Jackson JA, Kucheyev SO, Fang NX, Spadaccini CM. Ultralight, Ultrastiff Mechanical Metamaterials. Science. 2014;344:1373–7.
7. Pauly S, Lober L, Petters R, Stoica M, Scudino S, Kuhn U, Eckert J. Processing metallic glasses by selective laser melting. Mater Today. 2013;16:37–41.
8. Thamburaja P. Length scale effects on the shear localization process in metallic glasses: a theoretical and computational study. J Mech Phys Solids. 2011;59:1552–75.
9. Ekambaram R, Thamburaja P, Yang H, Li Y, Nikabdullah N. The multi-axial deformation behavior of bulk metallic glasses at high homologous temperatures. Int J Solids Struct. 2010;47:678–90.
10. Klusemann B, Bargmann S. Modeling and simulation of size effects in metallic glasses with a non-local continuum mechanics theory. J Mech Behav Mater. 2013;22:51–66.
11. Bargmann S, Xiao T, Klusemann B. Computational modelling of submicron-sized metallic glasses. Philos Mag. 2014;94:1–19.
12. Sarac B, Klusemann B, Xiao T, Bargmann S. Materials by design: An experimental and computational investigation on the microanatomy arrangement of porous metallic glasses. Acta Mater. 2014;77:411–22.
13. Sarac B, Wilmers J, Bargmann S. Property optimization of porous metallic glasses via structural design. Mater Lett. 2014;134:306–10.

Index

B. Sarac, *Microstructure-Property Optimization in Metallic Glasses*, Springer Theses,
DOI 10.1007/978-3-319-13033-0